FORGED BY FIRE
THE CULTURAL TENDING OF TREES AND FORESTS IN BIG SUR AND BEYOND

LEE KLINGER

FOREWORD BY
TOM LITTLE BEAR NASON

SUDDEN OAK LIFE PRESS

Copyright ©2024 by Lee Klinger
All rights reserved.

No part of this book may be reproduced in any form without prior permission in writing from the author.

Independently Published by Sudden Oak Life Press
Big Sur, California

ISBN: 9798322439752

Cover design by Yolanda Cazares.
Design consultation and book production by
Alexis Masters.
All photographs ©2024 Lee Klinger,
except when noted.
Back cover photo of Little Bear by
Michelle Magdalena Maddox.
Photo of Lee Klinger by Sonya Query.

To the Grandmothers and Grandfathers, both the Two-legged and the Standing Ones, who inspired this work.

CONTENTS

Foreword	1
Preface And Acknowledgments	5
Prologue	15
Chapter 1	21
The Tended Landscape and Traditional Ecological Knowledge	
Chapter 2	27
The Ancestor Oaks	
Chapter 3	47
The Redwood Giants	
Chapter 4	62
On the Age and Provenance of the Monterey Cypress	
Chapter 5	68
Ancient Trees and Shell Middens	
Chapter 6	79
The Cryptic Ecology of Mosses and Lichens	
Chapter 7	93
A Current Understanding of Forest Health and Fire Ecology	
Chapter 8	113
Returning the Gift: The Practice of Fire Mimicry	
Chapter 9	141
Assessing the Efficacy of Fire Mimicry	
Epilogue	160
Remembering Our Place in Nature	
Appendix	165
Species Nomenclature of Trees and Shrubs	
Bibliography	169
About The Author	187
Notes	189

FOREWORD

There are prayers and songs for the care of the ancient ones, the tree people, the stone people, all the old things. We pray and sing them often. We have songs for animals, we have songs for rocks, we have songs for springs, and all the beings which tie us together.

To me this book reads like a song for the trees. It's the first time I've been able to actually see our Tribe represented through the eyes of a person with a true grasp of Western science and Esselen culture. Since the moment, years ago, when we shared time together in a sacred redwood grove, each of us silently aware of the ancestral knowledge held in the trees, I have considered Lee Klinger as an Esselen, like me, with a different skin, but we're brothers.

After reading this book I've been able to grasp, even more.

Lee's mind, what he's experienced, everything he has learned, everything he has talked about, to the point of realizing that Lee and I speak the same language. We see things similarly, very much so. So as a brother, as an educated brother and fellow forester, in these ways, Lee has revealed in finer detail the essence of the Esselen and other Indigenous Peoples — the true, accurate voice of what we did.

I have worked all my life, as did my father and grandfather, to help our community understand the Esselen People and the importance of keeping fire on the land. I have talked with Cal Fire folks, other foresters, agents of the government, researchers, but the communications are often challenging. Lately I've been trying to work with a guy from the university, and we don't even speak the same language. I just tell him, I don't understand what you're talking about, because he's speaking in all these technical terms and using all these Latin names.

I don't speak those kinds of terms. I speak, thank you grandmother, and trees, and forest, and mushroom, and little grotto, and little waterfall, and pond, and pond turtle. The Esselen speak the language of the Great Mother.

It's sometimes difficult for me to talk about the loss of ancient forests in Big Sur, products of thousands of years of care by my People, all occurring in my lifetime, on my watch. The Ancestors cannot be pleased. I have seen old-growth ponderosa pines, incense cedars, Douglas firs, and Santa Lucia firs disappear from whole valleys because of damaging wildfires. My grandfather and my father warned the Forest Service repeatedly that this would happen unless fire was returned to the land.

You would think that after so many firestorms raging through Big Sur in recent decades, and all the lost homes and forests, that our community would try a different approach, one that is more open to fire, working with fire not against it. But fire is still being suppressed throughout most of our region. Between the many locals that understand the importance of healthy fire but don't have the means or knowledge to put fire back on their lands, to those who hold onto the wilderness ideology that any intervention by humans is a crime against nature, who is left to take the lead?

My dad tried to take the lead, and my grandfather, but few followed. Those of us who did are carrying on the fire practices learned and refined over hundreds and even thousands of years. I've been doing it for 63 years, 64 years — nearly since birth. Our fires were never large. Sometimes they would be thirteen acres, or forty acres, or four hundred acres. Probably the biggest one my dad ever did was about a thousand acres. But we burned *every* year.

Then in the 1970s the authorities stopped us from burning, writing us citations. Now look where we are at. In just twenty years we have lost the majority of our big trees, and many others are dying. The big grandfather oaks are dying because of all the little oaks growing like weeds around them. Redwoods too. Fire must be returned to the land and there are people like me and other Native brothers with that knowledge who are willing and able to do it.

Of course, no book can tell the whole story of the Esselen People and our connections to the Tree People. We have the rare albino redwood trees, with leaves of pure white, which grow in sacred places on the Esselen lands (*Figure* 1). Some of our big ponderosa pines, those few that remain on the ridgetops, have been culturally modified by my Ancestors to serve as marker trees for territories and trade routes. The great sycamores of the valley provide us with summer shade, and we have close relationships with the buckeye, the bay laurel, and the elderberry trees.

Figure 1. Sacred white redwood on Esselen land.

This book opens many doors that I hope others too will enter. But the door to fire mimicry, which Lee has entered, is the most relevant for today. The practice of fire mimicry is exactly what our People have done for thousands of years, we just never called it by that name. Nonetheless we would prune the trees and thin the forest with fire and axes, and fertilize the soils with ash and biochar from our fire pits and seashells from our middens. If we had chain saws back then, rather than stone axes, we would have used them too. There are a lot of forests to tend here in Big Sur!

This work tells the story of the Esselen connections with the land and trees better than anyone has done before. Several times I was brought to tears by the truth in these words. I am grateful for the information presented here and I hope that others will follow up on this relevant wisdom of our time.

A'ho.

Tom Little Bear Nason, *Tribal Chairman, Esselen Tribe of Monterey County* (March 2024)

PREFACE AND ACKNOWLEDGMENTS

When this world was finished, the eagle, the humming-bird, and Coyote were standing on the top of **Pico Blanco**. *When the water rose to their feet, the eagle, carrying the humming-bird and Coyote, flew to the* **Sierra de Gabilan**. *There they stood until the water went down. Then the eagle sent Coyote down the mountain to see if the world were dry. Coyote came back and said: "The whole world is dry." The eagle said to him: "Go and look in the river. See what there is there." Coyote came back and said: "There is a beautiful girl."* — Ohlone Opening lines of the Esselen Creation Story[1]

Gazing south from where I write these words is an iconic landscape of steep, vibrant hills covered in coastal prairie, chaparral, and clusters of evergreen trees. The view is bounded on the western horizon by an everchanging sky above and a living seascape of cresting waves and giant kelp forests below. To the east are the white, marble-capped peaks of the Santa Lucia Range, and in between are the sandy beaches at the mouth of the Little Sur River which gradually arc southwest to a massive rock jutting out into the Pacific Ocean — Point Sur. It is thought that the name of this area's original inhabitants, the Esselen, comes from their word *Ex'xien*, meaning "the rock," in reference to present-day Point Sur (*Figure* 2).[2] It is with gratitude and humility that I live on the unceded lands of the Esselen.

Figure 2. *Ex'xien* — "the rock", aka Point Sur.

The Esselen are an enigmatic tribe that was once reported, incorrectly, to be culturally extinct.[3] They have inhabited this small but diverse region of rugged mountains and jagged coasts centered around Big Sur for many thousands of years. Pico Blanco, Point Sur, Esalen hot springs, Tassajara hot springs, Arroyo Seco, and Soledad are all within their homelands.[4] The Esselen language (*Huelel*) has faded, but enough has been preserved to know that it was unusual and distinct among the California tribes. Based on linguistic evidence it is likely that the Esselen territory was once much more extensive, perhaps stretching from Santa Barbara County to the San Francisco Bay.[5]

Today's Esselen people are still a force shaping the face of Big

Sur. In 2020 the State of California helped to return to the Esselen Tribe of Monterey County 1200 acres of their homelands that include a stretch of the Little Sur River flowing at the base of **Pico Blanco**, the "Center of the Esselen World."[6] These lands are now being actively studied and managed by me and other members of the Department of Natural Resources of the Esselen Tribe. Recently awarded Cal Fire funds from a 2023 Tribal Wildfire Resilience Grant will ensure proper management of these newly acquired lands for several years to come.

I must also sincerely acknowledge the nearby Rumsen, Mutsun Ohlone-Costanoan, Amah Mutsun, Awaswas, Tamien, Ohlone, Salinan, Coast Miwok, and Pomo Tribes, as many of the observations and tending practices described in this book are reported from their Tribal lands too.

Yet, as my Native friends remind me, a land acknowledgement is but a hollow gesture if it does not involve some action in support of the Peoples and their stewardship of the land. For my part, this entire book reflects an effort to go beyond a land acknowledgement through acts of tending these ancestral lands and trees, because "returning the gift" seems like the right thing to do.[7]

I have not always lived in California, but I have always had a great love for and curiosity about the trees around me. During most of the 1970s I lived in a remote cabin at the base of Independence Mountain on the western slope of the Colorado Front Range. The cabin was situated next to Jones Gulch, a pristine source of year-round water running through a dense forest of Engelmann spruce, subalpine fir, lodgepole pine, and quaking aspen. Scattered throughout the forest were stumps of the trees I had felled to build the cabin. Other stumps were those of dead trees I had cut for cured firewood. After a few years, most of the dead trees and branches within view of the cabin had become ashes in my wood stove which I then used to fertilize the nearby aspen trees. The surrounding forest also contained many sickly-looking trees, a condition that I now realize was likely due to their high density. For the practical reasons of needing to heat my home and reduce the fire hazard around it, I thinned the forest of these suppressed trees year after year, and eventually began seeing the canopies of the remaining trees become lusher and denser. That is one reason why, forty-some years later, I find myself still doing these same practices — though in more thoughtful and efficient ways.

At that time, I had no knowledge that the Ute Indians, for millennia, had tended the local forests with cultural fires. None of the trees from their former habitation here remained. The oldest evidence of humans I could find were the abandoned excavations, mine shafts, and tailing piles from nineteenth-century gold and silver miners. I assumed (correctly) that most of the original forest had been logged and/or burned with ongoing exploitation of the land, creating the dense stands of young trees that dominated the hillsides.

Looking back at my education and professional life, I sometimes wonder how I navigated my way through this settler society. Raised in a white, midwestern American culture, my early education paid little attention to, or appreciation for, the Peoples of the Land. While local lore, place names, and the occasional discovery of an arrowhead engaged my imagination of the Native Erie Indians, they were treated as extinct beings in the Ohio history texts. Later, in my training as an ecologist at the University of Colorado, this lack of awareness of the scope and intensity of Native land practices became a major blind spot, leaving me ill-equipped to understand unusual patterns in forests and hillsides, things that could not be explained either by the impacts of modern humans or by the forces of nature alone. I lived in a fantasyland of John Muir and his wilderness ethos, and thus learned very well the language and stories of the colonizers. [8] And so for many years it simply did not occur to me that the Native Peoples played much of a role in shaping the so-called wilderness that I observed and studied.

My professional scientific training, while deficient in Native culture and knowledge, did provide me one gem of an idea that has greatly aided my understanding of Indigenous worldviews. In the late 1980s, I was introduced to Gaia theory, the notion that the earth is a living planetary system. And I soon met the late James Lovelock and the late Lynn Margulis, who together conceptualized Gaia theory as we know it today.[9] Deeply inspired, I took a position as a staff scientist at the National Center for Atmospheric Research (NCAR) in Boulder, Colorado and became fully involved in the science of Gaia. I attended all of the "Gaia in Oxford" meetings in the 1990s and have published several peer-reviewed research papers and book chapters on Gaia theory and geophysiology.[10]

NCAR is run by the University Corporation for Atmospheric Research (UCAR), which sponsors a program called SOARS (Significant Opportunities in Atmospheric Research and Science). SOARS seeks to involve students from groups that are historically underrepresented in the sciences.[11] As part of the program, students are paired with mentors from the NCAR scientific staff for ten-week summer internships. I was chosen to be a SOARS mentor for the inaugural class in 1996 and happily began interviewing candidates.

One student, a member of the Diné Tribe, was intrigued when I first described my work around Gaia theory. He immediately replied, in an elevated and proud voice, "My mother always told me the earth was alive!" Ever since, those words have echoed in me — and sometimes they still bring tears to my eyes.

Of course, I became his mentor. But all while he was mentoring me on what it is like to be raised with the belief that the earth is alive. As an objective scientist, I had to treat Gaia as a theory, not a belief. And a few years later I mentored another Native American student, a member of the Cherokee Tribe. Afterwards it all made me wonder: What would I have become had my mother, my father, and others in my extended family believed in, worshiped, and inhabited accordingly, a living planet?

In the early 2000s I moved to the Central Coast of California and began focusing my attention on the various ecological anomalies in the landscapes around Big Sur, including ancient cohorts of oddly shaped oak trees and peculiar groupings and skewed age structures in old-growth redwood groves. It quickly became apparent (*spoiler alert*) that these oddities were the result of two different cultural imprints juxtaposed on the landscape, the crowded and overgrown forests created by fire-suppression measures of Western colonial culture, and the many relic stands of ancient trees that originated under the care of the Esselen People.

Around this time, I also began reading the works of M. Kat Anderson, Dennis Martinez, Frank Lake, and Melissa Nelson to deepen my awareness of the worldviews, cultural ethos, and tending practices of the California Native Peoples. And, in 2010, I met Tom "Little Bear" Nason, Tribal Chairman of the Esselen, who has subsequently shared with me stories and places steeped in Esselen lore. These days he affectionately calls me "Little Tree" as we go about our frequent efforts of surveying, tending, and burning of Esselen lands. In 2019 I was also introduced to Kanyon Sayers-Roods and her mother Ann Marie Sayers, Mutsun Ohlone-Costanoan members of Indian Canyon Nation in the nearby **Sierra de Gabilan**, who live on the only federally designated Indian land between Sonoma and Santa Barbara in the Central Coast region of California. Kanyon and I have taught dozens of workshops together under the ancestor oaks on this sacred land. In our frequent conversations, Kanyon has helped reshape my perceptions and language around Native issues through her thoughtful teachings on cultural competency.

All the while I have been keeping close tabs on the ecological studies of California native forests, especially the role of historical and present-day fires, and have found much evidence in the current literature that the use of fire and other traditional tending practices by Indigenous Peoples were, and continue to be, sound and sensible. This has raised an important question: Could the recent spate of tree mortality outbreaks and catastrophic wildfires in California be tied to the fact that our forests are no longer being widely tended by the Native Peoples? This book is my attempt to address this question directly.

Modern-day California Tribes are well-aware of the human forces that have shaped these lands. Their ancestors persevered through unbelievable hardships to ensure that much of this knowledge has remained in the hands of the People. This is particularly true around the trove of information on cultural burning practices that have been preserved and, thankfully, are now being shared by various California Tribes, as will be described further in the Epilogue. I am indebted to all the wisdom provided to me by Indigenous Peoples and their allies.

This book is written for everyone, including tribal ecologists, land managers, and landowners who are, or will be, affected by forest decline and wildfires, as well as arborists, educators, scientists, students, and others who care about trees and the environment. My intent here is to present the relevant facts as accurately as possible while being aware, as a non-Native American, of the inherent limitations in my attempts to integrate these findings with Traditional Knowledge. I'm certain that some of what I say here will, as it should, be challenged both by Native Peoples and by the Western scientists of whom I write. Whatever criticisms are made, I trust they will lead to progress in our ability to better tend the forests.

Before proceeding, I must point out the difference in citational policies of Western academic versus Indigenous knowledge systems. Western science emphasizes generalizable knowledge that is then published, which means I can cite it in this book. Indigenous knowledge is passed down over generations and is culturally specific, place-based, even sacred. It may thus not be published and available for citation even though it is just as relevant and valid as that available through Western academic sources.

Throughout this book I have chosen to use English (Imperial) units rather than metric units as I suspect my intended audience will be more familiar with the former. Likewise, references to the common names of plants follow the colloquial (settler) usage, while the Latin (*Genus species*) plant names follow the Linnean classification system. The Latin taxonomy of the woody plants mentioned in the text by their common names is given in the Appendix: Species Nomenclature of Trees and Shrubs. When known, Esselen language (*Huelel*) names are included in italics within the text.

In my writing here about Indigenous Peoples, the use of terms and capitalizations follows the guidelines suggested in *Elements of Indigenous Style* by Gregory Younging.[12] I am keenly aware of how careless language can, even unintentionally, diminish and even erase the significance of Native Peoples and their cultures.

Among the many kind friends and colleagues who have helped in this work I extend sincere thanks to Tom Little Bear Nason, Cara Nason, Jana Nason, Chanel Keller, and Cliff Escobar of the Esselen Tribe of Monterey County; Kanyon Sayers-Roods and Ann Marie Sayers of Indian Canyon Nation; Ron Goode of the North Fork Mono Tribe; Leo Lauchere, the Gorski brothers Ero and Jake, and the entire EcoCamp Coyote team; David Shaw and John Valenzuela of Santa Cruz Permaculture; Neville Fay of Treeworks Environmental Practice (U.K.); Jared Childress of the Central Coast Prescribed Burn Association; Steve Davis (retired, but still active burn boss) of the U.S. Forest Service; Benjamin Eichorn of Central Coast Land Management; and the dedicated practitioners of fire mimicry who have worked diligently with me at Sudden Oak Life to get these results, including Rob Larson, Jorge Espinosa, Patrick Garretson, my lovely daughters Ava Klinger and Sonya Query, Jamie Self, Lauren Gamblin, Jeff "Willy" Wilson, Daniel Brooke, Rob Coleman, Martin Palafox, Jasmine Horan, the Ryan brothers Mark, Miles, and Jay, Chris Salem, TJ "Tree Jay" Lee, Ellen Katharine Rex, Colin Manor, Shannon Boyle, Ana Paula Teeple, Giovanna Piumarta, Sara S'Jegers, Aubrey Gates, David Abrahamson, David Yoshida, Christopher Vitale, Nicole Wong, Elena Staley, Rusty Sparks and too many others to be able to name them all here. I sincerely appreciate your help! Critical reviews by Madeline Sides, Cecil Frost, Stuart Abel, Miles Ryan, Lee Vierling, Oliver Tickell,

PREFACE AND ACKNOWLEDGMENTS

Shane Brown, and David Shearer have added significantly to the quality of this book. A detailed text editing and review were provided by David Moffat, and the cover design and photo editing are by Yolanda Cázares, with assistance from Kelly Smith Cassidy and Eira Mooney. Alexis Masters also helped with final design and formatting of the manuscript. Finally, I wish to thank the many landowners and property managers who trusted us with the care of your trees and forests. This work could not have been possible without your interest and support.

PROLOGUE

Late one December night in Big Sur, nestled in my yurt beneath a grove of young redwood trees, I was stirred from a deep sleep by an urge to pee. I rose to check the time and found it was around 3 AM. Gusty winds were slapping against the fabric roof as I skirted past the bathroom and stepped out onto the deck to get a breath of fresh air while fertilizing the redwoods. Though I often opted to pee off my deck, I sometimes wonder what would have happened that night had I used my indoor bathroom instead.

Upon reaching the deck, I immediately saw a red glow in the eastern sky. A brief inkling that it was the early light of sunrise quickly faded as I realized that it was the middle of the night, the mountain I lived on was on fire, and the Santa Ana winds were blowing strong.

I peed.

I then got dressed, grabbed my headlamp, and ran to the neighbors to alert them, but found they had already left. Returning to my yurt I did something that may seem odd if you're not a coffee drinker. I brewed a strong coffee and, sitting calmly, slowly savored the entire cup, all the while coming up with a fire plan.

Once caffeinated, I proceeded to take photos of my library and of everything else in my home. I then gathered what important items I could fit in my VW van and tried to drive out the only exit road. But I soon found that the road was already engulfed in flames. So I turned around and parked my van near a storage shed where I had previously cleared a fire break. I then began to fight the fire with a shovel as it burned downslope through the chaparral, hoping to use my driveway as a fire line. After about half an hour of successfully keeping the fire from crossing the road I noticed, however, that off to the south the fire had flanked me and was now burning upslope through a dense redwood forest towards my yurt (*Figure* 3). It was time get the hell out of there!

Turning to my contingency exit plan, I squeezed all that was necessary into a backpack and, in the light of my headlamp, located a trail I had cut a few years back and followed it down a steep side canyon and up to an adjacent ridge, away from the fire. When I finally reached a vantage point on the ridge, I stopped for a bit and, in the early morning glow, watched cathartically as my home spirited up in flames.

I spent most of the next two days helping my neighbors clear defensible space. And through our common effort, the fire was finally contained. In the end, the Pfeiffer fire destroyed 34 homes and partially destroyed the properties of many other residents. Thankfully, no one was killed or injured.[1]

Afterwards, returning to the charred remains of my yurt, I found that Sylvia (my VW van) and storage area had survived the fire with minor damage. Fortunately, I did have adequate fire clearance in some places. I would later find out that around midnight on December 16, 2013, soon after the fire started, a sheriff's deputy had driven down the road near my yurt and announced an evacuation order over his loudspeaker. My elderly neighbors heard it and fled, but I sleep soundly and with my good ear on the pillow I did not hear a thing.

So here we are, a decade later, and California is still on fire. The recent years have seen the largest and deadliest wildfires on record. Not only are so many lives and homes being lost, but vast areas of forest are also being destroyed. Frequent evacuations disrupt our lives, as do the PG&E power shutdowns during red flag events, and there are extended periods when we are forced to breathe toxic air. Even after the fires are out, our roadways and other service infrastructures are subject to damage from landslides and mudflows emanating from heavily burnt hillsides. Water sources are likewise being contaminated with caustic substances produced by burned infrastructures. Just last year I was forced to evacuate when a wildfire burning in the middle of winter (January) came within a mile of my new home here in Big Sur!

Figure 3. The last photo of my home while fleeing a Big Sur wildfire on December 16, 2013.

While living in Big Sur I have witnessed more than a dozen

wildfires, most notably the Basin Complex fire in 2008, the Pfeiffer fire in 2013, the Sobranes fire in 2016,[2] and the Dolan fire in 2020 (*Figure* 4). The immediate and subsequent impact of each of these fires has disrupted the entire community in various ways, from the loss of homes and property to roadway failures that isolate people and businesses. To the credit of our strong and resilient citizens, we have ended up sharing our time and resources, which has allowed us to survive these challenges. Still, iconic forests have been lost and others are so fire damaged that, even after a decade, popular trails remain closed to visitors on certain Big Sur state park and U.S. Forest Service lands. In the aftermath of these fires, we're also seeing so much partially burnt and unburnt fuel remaining, that the same areas end up burning fiercely again in subsequent fires.

Figure 4. Basin Complex fire above Big Sur, June 2008.

It may seem that wildfires now are rampant, but we also know that fire has been a part of the Big Sur landscape for millennia. As the Esselen People inform us, life here is all about forging a covenant with fire. Most of the responsible land managers that I know in Big Sur follow this covenant by spending their time starting (legal) fires during the burn season, controlling wildfires in the dry season, and reducing fuels in preparation for the next fire that is certain to come.

In the words of Tom Little Bear Nason (*Figure* 5), Tribal Chairman of the Esselen Tribe of Monterey County:

> Our family has lived, loved and shared this sacred land for many generations and we always will forever. Our family… practice[ed] traditional Native indigenous Esselen tribal burning of this valley up until 1970s. [Then the] government said STOP BURNING!! My Forefathers all told them that by ordering… [fire] off these lands it would begin a dangerous situation by allowing the brush and scrub to grow out control, [so that] the forests would become choked… and when a natural force like lightning comes it would cause the lower brush to burn at high heat and kill the trees. In this photo [see endnote for link] you see many big ponderosa pine trees and open meadows surrounding them. As Natives of this land, we knew how to manage our lands and the forests. Since the 1940s, we've had many wildfires come through the Santa Lucia Mountains, and some were good for the land, [but] most have been extremely damaging. My family and tribe have seen our beloved and sacred places here in Big Sur changing so much due to imbalance and [we feel] deep sadness for losing so many of the old trees. We need change and it's very challenging for all of us to live with so many fires so frequently!! Prayers and Respect to all who listen to Mother Nature.[3]

Figure 5. Tom Little Bear Nason (photo by Michelle Magdalena Maddox 2022).

1

THE TENDED LANDSCAPE AND TRADITIONAL ECOLOGICAL KNOWLEDGE

After half a lifetime examining the "natural" world through the lenses of botany, ecology, chemistry, and geophysiology I now find myself exploring a terrain beyond the theories and quantifiable facts of Western science. This terrain extends into the realm of Traditional Ecological Knowledge (TEK), where nature is seen as a process inseparable from the People of the Land. The Indigenous author and scientist Robin Wall Kimmerer, in her book *Gathering Moss*, succinctly describes the territory I seem to have entered:

> In Indigenous ways of knowing, we say that a thing cannot be understood until it is known by all four aspects of our being: mind, body, emotion, and spirit. The scientific way of knowing relies only on empirical information from the world, gathered by the body and interpreted by the mind. In order to tell the mosses' story I need both approaches, objective and subjective.[1]

These two worlds are not easily traversed, but it is imperative that we try. As Gregory Cajete, director of the Native American

Studies program at the University of New Mexico, writes "Native and Western cultures, in their seemingly irreconcilably different ways of knowing and relating to the natural world, must search for common ground and a basis for dialogue."[2]

The new wealth of information to be provided by Western ecological studies imbued with local TEK can allow for greater insight into the place of humans in nature.[3] In an effort to obtain objective knowledge of a natural phenomenon, the Western scientist typically aspires to be a removed, unbiased observer in gathering generalizable facts and data. Yet, while this approach has clearly provided a significant quantitative understanding of the natural world, it is lacking when it comes to elaborating the human roles and best practices for living in and restoring most ecosystems. Western science also tends to negate a part of our humanity — the fact that we exist not just from our objective knowledge, but through our, and our ancestors', countless interactions with nature. This, I believe, is where TEK can best inform Western science and restoration ecology.[4]

TEK is broadly defined as those unique bodies of Indigenous knowledge, beliefs, and practices that are acquired and passed on over many generations in order to properly manage the traditional natural resources of a given place. The fated blending of TEK and Western science has already begun and is generally being well-received. A stellar example of this is Robin Wall Kimmerer's award-winning book *Braiding Sweetgrass*, which, for me, has highlighted the deep wisdom about the world that can be revealed by examining it through the lenses of both scientific and Indigenous knowledge.[5]

Dennis Martinez, a renowned restoration ecologist and Indigenous rights leader, points out that all of humanity shares a common evolutionary cultural heritage. He describes this as: "an ancient way of being with nature, not only with plants and animals, but with the primal natural forces of fire, water, winds, and the earth — a way of relating respectfully to all life as kin, and the earth as a nurturing mother."[6] Martinez also refers to the suite of Indigenous cultural land-care practices (i.e., TEK) as forming a "kincentric" system, creating ecologies that arise from long-standing reciprocal relationships in cultural landscapes and seascapes.

Melissa Nelson, a professor of American Indian Studies at San Francisco State University, also writes of TEK: "The Indigenous ethics of TEK are longstanding, intergenerational, and tied more to manner and behaviors than principles, in the sense that they are more about actions and hands-on activities than beliefs or ideals."[7] Indeed, most of my own direct interactions with California Native communities have involved hands-on stewardship, storytelling, and educational gatherings on their lands.

Tending the Wild, by M Kat Anderson, provides a further detailed account on the multitude of traditional tending practices employed by Indigenous California Tribes.[8] Her book considers evidence from the historical literature, archeological studies, fire history reconstructions, ecological analyses, and direct oral accounts of the tending customs of past and present-day California Indians. This work has helped strengthen my appreciation for, and broaden my understanding of, the scope of influence that Native Peoples have had on the landscapes of California.

At the core of Indigenous approaches to managing their lands is the concept of "reciprocity." Reciprocity, in this perspective, involves both material and spiritual acts of gratitude. Robin Wall Kimmerer refers to reciprocity as "returning the gift" and she points out that this is not merely an act of appreciation, but the way nature works![9] Thus, ecosystems can only function properly when a multitude of reciprocal processes involving production, consumption, and decomposition are all happening and in balance.

Supreme among the skills in traditional resource management was, and still is, the wise use of fire. This involves knowing precisely when, where, and how frequently to apply fire to a given ecosystem to obtain the desired results. The tool of cultural fire can be used for a suite of purposes such as improving food harvests, thinning overcrowded forests, clearing for habitation, promoting the regeneration of basketry materials, controlling pests and diseases, felling trees, and hunting game.[10] These and other applications of fire have resulted, over millennia, in the culturally modified landscapes of prairies, woodland savannas, and old-growth forests once thought to span all of North, Central, and South America.[11]

In Big Sur the Indigenous ethos practiced by the Esselen have created a rich mosaic of plant communities. Coastal prairie, coastal chaparral, oak savanna, forests of bay laurel, riparian forests, old-growth oak forests, old-growth redwood forests, and more can all still be found here in abundance within a single watershed (*Figure* 6).

Figure 6. Coast Ridge and Post Creek watershed in Big Sur.

I wish to be careful not to over-romanticize Indigenous cultures in their efforts to tend the land. Mistakes were certainly made, some of which may have led to the collapse of some Native populations, and perhaps even other species. However, Indigenous Peoples had time on their side, allowing for gradual adaptation to instances of mismanagement, enabling the eventual persistence of later generations.

Now, in California and widely throughout western North America, the general exclusion of cultural fires over the past two hundred years has had a profound impact on the native ecosystems. Many of the mature oak, pine, and redwood forests are now in decline, having become overcrowded and increasingly susceptible to rising temperatures, drought, disease, and catastrophic wildfire. Species-rich prairies are shrinking with the invasion of shrubs and young trees. Even aquatic ecosystems are struggling, with perennial streams drying up in the summers as the increasing number of trees in the watershed take up moisture from the ground. Fire suppression also affects the fertility of the soil, the expansion of non-native species, and the growth of mosses and lichens.

While there are various wide-ranging effects of fire on ecosystems, in the following chapters I will center my discussion on the consequences of cultural fire suppression on the old-growth oak and redwood forests of Central California, their stand structures, the underlying pH and fertility of soils, and the impacts of mosses and lichens. I will also discuss the role of Native shell middens at prominent village/gathering sites that once provided the basis for highly fertile soils, out of which grew (and still grow) ancient, culturally modified trees.

Most of the material in this book is presented with one eye on the problem of forest health and the other on possible solutions. It is clear to me that any large-scale solution will require a reintroduction of fire onto the landscape. Still, there are other means to help improve the health of native forests without putting fire to the land. I call this approach "fire mimicry" and describe, in the final three chapters, the TEK, science, methods, and efficacy of addressing the health of trees and soils using tools and materials other than fire. The practice of fire mimicry draws on information derived from ecological findings, soil analyses, archeological remains, historical reports and photos, and written and oral accounts of Indigenous Peoples in California and elsewhere.

Let me add that the information and interpretations I share here are not fixed. My understanding of the cultural ecology of California forests is fluid, changing with every visit to the forest and every lesson from a tree-wise person. As I discuss in the Epilogue, many gaps remain in both our scientific and Indigenous knowledge of forest management, and I'm hoping this work will inspire others to help fill these gaps.

2

THE ANCESTOR OAKS

The human history of Big Sur began many millennia ago when the ancestors of the Esselen People first appeared and began tending the land, trees, and seashores. Previously the ecosystems of the California Central Coast, according to paleoecological records, had been largely comprised of forests dominated by conifers.[1] These forests were subject to what I call a boom-or-bust ecology, which included lengthy disturbance-free periods of forest growth and development followed by occasional dry-season lightning strikes resulting in catastrophic wildfires.

Upon the arrival of the First Peoples, the coniferous forest ecosystems quickly transformed to, and have since remained, oak-dominated forests and woodland savannas, with little evidence of severe wildfires.[2] Thus, when the first Western colonizers, the Spanish Vizcaino Expedition, landed in this region in 1602, they described the land as "fertile" and dominated by large oaks and tall pines. They likewise recorded that there was much wild game and countless species of birds, and that the area was thickly populated by Indians, who were "a gentle and peaceable people" subsisting mainly on seafood and acorns.[3]

Besides these noteworthy first impressions, Vizcaino described how, immediately upon his arrival at Monterey Bay, his party said a mass at an altar they set up beneath "a giant oak" near the shore.[4] Over a century and a half later, in June of 1770, when Father Junipero Serra arrived in the area, he reported setting up an altar beneath this same giant oak, and saying a mass in commemoration of the first mass of 1602. And it was eventually recorded that this iconic "Vizcaino oak," a coast live oak, lived until 1904, perishing only after suffering severe decline due to faulty drainage from the construction of a nearby railroad.[5]

The above chronology of the Vizcaino oak has important bearing on the discussion that follows, as it points to the great longevity of oaks in this region. This oak, which was notable to Vizcaino not only for its size but its easy access from the shore, must have been just the first of countless giant oaks the expedition encountered on the land. Let us assume that in 1602 this "giant oak" was no less than two hundred years old, which I believe to be a conservative estimate. This would mean that upon its demise in 1904, it would have been at least five hundred years old, and probably much older. Had this oak not succumbed to the impacts of colonization, who knows, maybe the Vizcaino oak would still be alive!

Fortunately today, scattered throughout the rugged hills of Big Sur, can still be found thousands of stately groves of giant coast live oaks, canyon live oaks, black oaks, and tanbark oaks, as well as stands of valley oaks and blue oaks that are centuries old and date from a time when the Esselen People were the sole human occupants of the land (*Figure* 7).[6] The largest oak I have seen in Big Sur is a canyon live oak nearly ten feet in diameter, which, based on a tree ring count from a partial core, is estimated to be around eight hundred years in age. This and the many other large oaks in this region are well-spaced, with broad canopies and stout limbs that extend more-or-less horizontally, so that the width of the canopy is often greater than the tree is tall. It is clear from their branching

CHAPTER 2

patterns that these oaks developed for many centuries under open, park-like conditions, with little or no canopy encroachment from neighboring trees.

Figure 7. Archetypal California oak savanna on Esselen land.

Nowadays, however, growing under and around many of the old oaks are dense cohorts of young oaks, as well as bay laurels and, in places, young redwoods. The oldest of these newcomers are of an age contemporaneous with the arrival of colonial cultures and the intentional, and often violent, displacement of the Esselen Indians in Big Sur. Other trees are even younger, having established themselves in abundance when livestock grazing ended on various tracts of rangelands in recent decades. All of this is the result of the colonial suppression of cultural fires previously set by the Esselen to maintain the health of the oak groves.

. . .

Cultural dimorphism in California oaks and other native trees

While spending time, as ecologists are inclined to do, documenting the species and sizes of the trees, and examining their shapes, spacing, and habitats, I've come to see a distinct pattern in these groves. There is a characteristic dimorphism (two forms) evident in the branching habits of larger/older oaks vs. the smaller/younger oaks. The larger oaks, which I believe are 250+ years old, have conspicuous shapes involving multiple (three to six) major branches splaying outward from near the base of the trunk (*Figures* 8 and 9). By contrast, the adjacent smaller oaks display tall, erect leaders and a dichotomous branching morphology typical of young oaks. (*Figure* 10). This pattern is repeated grove after grove here in Big Sur, and the same is true in nearly every other location I've examined in California exhibiting a similar juxtaposition of young and old trees, regardless of the species.

Figure 8. A culturally modified canyon live oak on Esselen land, estimated to be around eight hundred years old based on a partial tree ring sample (photo by Rob Larson).

CHAPTER 2

Figure 9. A culturally modified coast live oak, estimated at more than three hundred years old, growing in open conditions on Rumsen land. Note the broad, spreading canopy that is wider than the tree is tall.

Figure 10. A young, untended coast live oak, less than one hundred years old, growing in open conditions on Rumsen land near the previously shown culturally modified oak (*Figure* 9). Note the dichotomous branching pattern and tall, erect form typical of most oaks.

So what is the origin of this widespread dimorphism in oak

cohorts? Given that the establishment and growth of older oaks occurred when the Esselen occupied the land, while the younger oaks are contemporaneous with Western culture, I can come to only one conclusion: that the distinctive forms of the older oaks are the result of pruning, pollarding, and other tending practices by many generations of Esselen families. I call these ancient, culturally modified oaks, "ancestor" trees, and I consider them to be living artifacts of Esselen Indian culture.[7]

Today the pollarding of orchard trees is a common agricultural practice that helps maximize the productivity and ease of collection of fruits and nuts. It involves the repeated pruning of the upper leader branches so as to encourage multiple lateral branches that eventually splay outward from near the base of the tree. This practice can also extend the life of a tree by lessening its top-heaviness and making it more wind resistant.[8] As my colleague Neville Fey, who has shown me numerous ancient pollarded oaks near his home in southwest England, writes, "A pollard literally means 'beheading' and the tradition originated in cutting trees above the height that animals graze."[9]

In Europe, the pollarding of oaks is an age-old custom that is still practiced in places like the U.K.,[10] France,[11] Spain,[12] and Turkey.[13] Archaeological excavations in Europe have uncovered pollarded trees dating to the Iron Age, and a fossil pollarded oak found in the U.K. has been dated even earlier, to 3,400 years ago.[14]

The TEK literature describes how the broad, spreading shapes of the large, old-growth oaks are the result of burning to create open areas for the trees to branch laterally. Burning around the oaks involves setting ground fires during the appropriate season every few years.[15] This process keeps the areas around them free of competing trees and shrubs and creates open-canopy, savanna-like habitats. Without encroachment by the canopies of neighboring trees, oaks are able to grow laterally, creating an optimal shape for acorn agriculture. Cultural burning together with the frequent knocking by acorn gatherers using long sticks, thus effectively pruned the oaks by removing smaller limbs and promoting lateral branches, resulted in these pollarded forms.[16]

The practices of pollarding and coppicing (cutting plants close to or at their bases to create multiple stems) are often confused and used interchangeably. TEK writings frequently mention the pollarding, coppicing, and pruning of trees and other plants, although this nomenclature can have different meanings across the various cultures.

Upon noticing this pronounced variance in forms of older vs. younger trees it is hard to un-notice it. The pattern is apparent in groves of coast live oaks, canyon live oaks, valley oaks, blue oaks, black oaks, and tanbark oaks, as well as Oregon white oaks, interior live oaks, and Engelmann oaks throughout their ranges (*Figure* 11). Indeed, I have been told repeatedly by people to whom I've shown this to, that they now see the local forests "totally differently."

Figure 11. Culturally modified valley oak surrounded by younger erect-growing valley oaks on Pomo land.

CHAPTER 2

Not only were California's oaks tended this way. Many old-growth Monterey pines (*Figure* 12), California bay laurels (*Figure* 13), California buckeyes (*Figure* 14), Pacific madrones (*Figure* 15), and California sycamores bear pollarded forms. In my mind these trees are living cultural artifacts of the local Native Peoples, which stand as modern-day legacies to their ecological competency.

Figure 12. Culturally modified Monterey pine on Rumsen land.

Figure 13. Culturally modified bay laurel on Ohlone land.

CHAPTER 2

Figure 14. Culturally modified California buckeye on Rumsen land.

Figure 15. Culturally modified Pacific madrone on Esselen land.

In the TEK literature pollarded trees fall into the category of culturally modified trees (CMTs). CMTs are a phenomenon of forest-dwelling peoples worldwide. Turner et al. (2009) describes CMTs as "Living trees from which materials are harvested (edible inner bark, pitch and resin, bark, branches), or which are modified through coppicing and pollarding to produce wood of a certain size and quality, or which are marked in some way for purposes of art, ceremony, or to indicate boundary lines or trails, all represent the potential of sustainable use and management of trees and forested regions."[17]

Tom Little Bear Nason tells me that the ancient, pollarded oaks were tended by generations of Native families. Women often took the lead in caring for and burning around the oaks, as well as gathering and processing the acorns. Families here and elsewhere used long sticks to knock the branches to harvest the acorns. I sometimes envision the Esselen children climbing the splayed branches and shaking them until the acorns fell onto the newly burnt ground, or perhaps onto woven mats laid beneath, to then be gathered. The low, broad forms of the tended oaks greatly facilitated the ease and safety of acorn harvesting.

Oaks in California Native cultures

In their book *The Esselen Indians of the Big Sur Country*, Gary Breschini and Trudy Haversat repeatedly reference the importance of oaks and their acorns in the lives of the Esselen. With regards to the oak resources of the Esselen, they quote Helen McCarthy:

> Acorns from all of these (native oak) species were used in the production of the staple food of California Indian peoples, including the Esselen. They were collected in the fall and dried and stored in large quantities, for both immediate and future use. The acorns were pounded into a fine flour and which is leached and cooked. The collection of acorns and management of their storage was essential to Esselen survival, and a focus of Esselen ceremony to thank the Creator for the gift of acorns/food and to encourage the continuing well being of the crop. Thus, these resources, the living trees, are central to Esselen heritage. In addition to the use of acorns as food, Esselen made necklaces of the acorns, sometime alternating acorns and manzanita berries on a string.[18]

It makes perfect sense that the Esselen would tend the oaks (*has*),[19] which were (and still are) a cultural keystone species in the local ecosystem.[20] And the abundance of acorns (*palatsa*) that the oaks produced served not only as a food source for the People, but also fed large populations of deer, bear, mice, squirrels, jays, and woodpeckers.[21] Among the oaks, too, were (and are) many edible fungi, including the prized chanterelle. The bark of the oaks also contains medicine; its extracts used to tan hides; and the dense, long-burning wood of the oaks has cooked many a meal and kept families warm for generations. Moreover, in the summer, what could be more appealing than gathering for story time under the cool shade of an ancient oak.

M. Kat Anderson elaborates on the central importance of oaks for Native California Tribes noting that, in addition to the common uses described above, oaks were also used to make basketry, utensils, digging sticks, battle armor, house infrastructure, and tinder for fire kits, as well as for dyes, games, and toys.[22] Native Californians also crafted various remedies from oaks, including making tea from the bark for bad coughs, and they used oak galls to make ink and to treat sores and toothaches.[23]

Besides oaks, there are many species of plants associated with oaks were (and are) also useful to the California Tribes. In *Secrets of the Oak Woodland*, Kate Marianchild describes the various Native uses of oak woodland affiliates such as toyon and poison oak, both of which can be utilized to make tools, teas, and tonics.[24]

The remarkable diversity of oaks in California, nearly twenty species and as many varieties, is likely related to the selection, cultivation, and subsequent speciation of oaks over thousands of years as a result of tending by the Native Peoples. Moreover, just to the south, in Mexico, there are in excess of 150 species of oaks.[25] Based on this unusually high species richness, it seems likely that California and Mexico were epicenters of oak domestication in the Americas. In this sense domestication was really a form of reciprocity. Both the oaks and the Peoples benefitted from this kincentric relationship.

It may be, as some have said, that the Native Peoples left the planting of most oaks to forgotten caches of acorns hidden away by the jays and squirrels. However, there are also reports of intentional planting of oaks. Florence Shipek states that the Kumeyaay of southern California planted oak trees, and that ancient oaks there are grouped around historically important Indigenous habitation sites.[26] This is a pattern that I, too, have observed across California, where groves of ancient oaks near known village sites exhibit circular or oval arrangements with a central open area in the middle. There is even evidence that the Ohlone planted oaks along the "Avenue of the Sun" on Mt. Hamilton so as to mark the position of the rising sun on the summer solstice.[27]

Oaks, too, have been a keystone species in my own life as a forest restoration ecologist. The majority of people who contact me are experiencing health issues with their oaks, especially groves of coast live oaks. Tending oaks has brought comfort to many of my clients and has provided me with a fair income, as well as a considerable wealth of information.

I also try to make the best of my failures. There are times when I must to deal with oaks that have succumb to disease and insect pests, including some I've tried to save. But these dead oaks need not go to waste; many of them can be cut into high-quality firewood. Living off the grid and using firewood to heat my home for most of my adult life, I have found that there is no wood in the western U.S. with such a high heat value as seasoned coast live oak.[28] And I regularly remove the ashes and bits of biochar from my stove to incorporate into the mineral compounds I use to fertilize other oaks.

I invite others to make oaks a keystone species in their own lives. You will not regret it!

The chemical ecology of oak forests

Along with their role in sustaining high productivity and species diversity in forest ecosystems, oaks also affect the chemical ecology of the air. Oak leaves emit, in significant quantities, dozens of hydrocarbons including isoprene, monoterpenes, sesquiterpenes, and organic acids. Together, these biogenic volatile organic compounds (VOC) and their oxidation products play important roles in the chemistry and physics of the forest environment (*Figure* 16).

Figure 16. Biogenic aerosol haze produced by hydrocarbon emissions from oak-dominated forests in the Great Smoky Mountains (Frank Kehren/Flickr).

Nearly all oaks emit isoprene (C_5H_8), a highly volatile and reactive light hydrocarbon that is produced in the chloroplasts of the leaves and released into the atmosphere. Is isoprene perhaps merely a waste product resulting from a metabolic inefficiency in the leaf cells? Clearly not, as oaks and other isoprene emitting plants also manufacture an enzyme called isoprene synthase that converts organic precursors into isoprene. Hence, isoprene is intentionally produced by plants, often in large amounts. Indeed, in an oak forest the carbon emitted as isoprene can represent one to two percent of the total annual net primary productivity.[29] For this amount of energy investment, isoprene must serve some adaptive function(s) for the oaks. And, given that most of the isoprene produced in the chloroplasts of the leaves escapes into the air, perhaps the forest atmosphere is where we should focus our inquiries into why oaks and other trees emit so much isoprene.

Once airborne, most of the isoprene molecules quickly undergo photochemical oxidation. The oxidation products of isoprene, such as methacrolein and methyl vinyl ketone, are precursors of cloud condensation nuclei (CCN). As the name suggests, these airborne biogenic particles enhance the formation of aerosols and clouds. So why would plants, like oaks, want to enhance formation of aerosol hazes and clouds, which would reduce direct sunlight, thus affecting photosynthesis? Perhaps because the aerosols reflect sunlight and cool the atmosphere, thus reducing heat stress in trees, and perhaps because clouds often produce fog and rainfall, providing the water that is essential in photosynthesis. The formation of CCN via biogenic VOC oxidation may thus be one of the main ways some forest ecosystems regulate atmospheric temperature and precipitation.

Furthermore, there appears to be another atmospheric feedback mechanism in forests involving isoprene. In the late 1990s while conducting research in Africa with a team of U.S., French, and Congolese scientists on the chemical ecology of the Congo rainforests,[30] I brought along a doctoral student, Lee Vierling, who worked with me at NCAR.[31] In his dissertation research on the isoprene emissions and light levels at various heights in the Congo rainforest canopy he found, surprisingly, that the tropical rainforest photosynthetic carbon assimilation is highest during partly cloudy conditions. This was due to a combination of decreased heat stress in the upper canopy (lowering respiratory carbon losses) and to increased photon flux from diffuse and reflected radiation coming from clouds.[32]

In other words, a perfectly clear sky day allows for less tropical rainforest photosynthetic carbon gain than one with some clouds. The reason is that clouds not only decrease the thermal load on the canopy leaves, they also disperse the sun's rays, allowing multiple angles of light to enter the forest subcanopy, enhancing photosynthesis on understory leaves that would otherwise receive low levels of direct sunlight.[33] This regulation of light and temperature in a forest canopy via isoprene oxidation and CCN production would, however, only make sense if isoprene emissions were highly dependent upon the rate of photosynthesis — which they are. Hence, at night when the canopy temperatures are lower (less respiratory carbon loss) and there is no sunlight, photosynthesis pauses and the leaves stop emitting isoprene — perhaps because there is no need to regulate the light and temperature then.

In this circumstance we don't yet know the percentage of cloud forming CCN coming from isoprene emissions by the local rainforest compared to exogenous sources of other biogenic particles, smoke, etc. But, hopefully, similar studies in different kinds of forests, including oak-dominated forests, are on the horizon.

Talking trees

Having measured, on numerous occasions, hundreds of chemical constituents emitted from oaks, it seems to me likely that at least some of these reactive chemicals also serve as airborne flows of ecological information.[34] Innovative research in this realm has already identified various gaseous biogenic hydrocarbon compounds released by trees subjected to insect herbivory, which appear to signal other nearby trees to bolster their defenses against insect attack.[35] It has even been suggested that chemical signaling in oaks may be involved in masting behavior, certain years when oaks over large regions synchronously produce surplus amounts of acorns.[36] This masting strategy is thought to help ensure reproductive success in the face of heavy acorn predation by mammals, birds, and insects.

The idea of ecosystem-level communication has been promoted in the recent works of Susanne Simard (*Finding the Mother Tree*)[37] and Peter Wohlleben (*The Hidden Life of Trees*),[38] who propose that trees communicate and transfer resources via common mycorrhizal networks (CMNs) comprised of fungal hyphae that connect their roots — popularly known as the "Wood Wide Web." Simard's findings suggest trees regularly exchange information and resources, and that the net flow of resources tends to be from mature ("mother") trees to younger trees. If true, this is a critically important finding, especially when assessing whether-or-not the health of a mother tree can become compromised by an overabundance of "nurse" trees. From what I've seen, old-growth oaks and other native trees (i.e., mother trees) often show signs of stress in densely wooded forests.

However, a recent review of the CMN literature provides a cautionary tale on the over hyping of these so-called Wood Wide Web results. Specifically, a study by J. Karst et al. reports evidence of a bias towards citing mainly positive effects of CMNs, and that no field studies have shown that mother trees preferentially send resources to nearby young trees.[39] Taking all this into account, I think that communication and resource exchange among oaks and other trees is quite likely, but the means and modes by which this happens, whether via the atmosphere and/or the soils, remains largely unknown at this point.

As an aside, I am sometimes asked by friends if I can hear trees "talk." While I'm deeply inspired in the presence of trees, I must say that I do not hear them talk, at least not in the sense of consciously receiving words or thoughts. BUT they do tell me stories, however slowly, over months and years, with a kind of "sign language" that becomes apparent with each return visit and offering to a particular tree or grove. The words of this sign language are simply the visual cues a tree displays: the density of its canopy, the lushness of its foliage, the quality of its bark, and the healing of its wounds. Every time I return with a gift to a tree I know, an oak or otherwise, each of these qualities differs and can be observed and documented photographically. Thus, all the trees that I visit year-after-year communicate by revealing their transformation over time. Images of this visual language are presented in Chapter 9.

As I have thus tried to describe in this chapter, a blending of scientific information and Indigenous knowledge reveals how in California, and elsewhere in the Americas, a keystone plant species (oaks) interacted for thousands of years with a keystone animal species (humans) to create a vast kincentric ecosystem. In this ecosystem, the oaks provided sustenance, tools, fuel, etc., while the humans reciprocated by applying good fire and other tending practices, resulting in healthy trees and an abundance of acorns and other byproducts that benefitted countless plant, animal, and fungal species.

3
THE REDWOOD GIANTS

There are forests of giant redwoods (*s`u'men*) in Big Sur that hold memories of times, millennia ago, when the big trees grew taller and more abundantly throughout these lands. The ancient redwoods here comprise the southernmost present-day extent of native redwoods along the western coast of North America. They have adapted to the seasonally dry conditions by harvesting water from fog with their needle-like leaves. The immense longevity of coast redwoods is attributed, in part, to their remarkable resistance to disease and insect pests. Rarely are pathogens or insect infestations found to be the proximal cause of redwood decline. Redwood bark is also highly effective in protecting the trunks and branches from fire damage, allowing the trees to better survive wildfires.

With the arrival of Western settlers, timber harvesting and land development in Big Sur have spelled the fate of too many of these magnificent trees. Still, more than a few have escaped the axe thanks to Big Sur's rugged terrain, which created impassable barriers to logging operations. The oldest remaining primary (unlogged) redwood groves are scattered along remote valley bottoms and in steep ravines, but some can be found in isolated stands high atop grassy ridges overlooking the coast (*Figure* 17).

Figure 17. Ancient redwood grove atop a coastal ridge on Esselen Land. A slash pile is burning on the foreground.

Coast redwood forests are known to support the highest species diversity of any North American temperate forest. In the canopies alone can be found thousands of species of insects and hundreds of different epiphytic mosses, lichens, fungi, ferns, shrubs, and even small trees. Many of these species are unique to the upper branches of redwoods, which, at a height of more than two hundred feet, create habitats and microclimates rarely found in other forest environments.[1] The fire-carved cavities in the base of large redwoods also serve as excellent bat habitats, and a higher diversity of both bats and birds is recorded among these fire-scarred "legacy" trees compared to groves of younger, unscarred redwoods.[2]

Redwood circles

The largest redwoods are typically grouped in cathedral-like groves of up to ten trees, sometimes arranged in circular or semicircular patterns (*Figure* 18). The circular shapes of redwood groves, commonly called "fairy rings," is thought to arise from the adventitious sprouting of young trees around the base of a current or former "parent" tree. Indeed, I have seen numerous small circles (ten to fifteen feet in diameter) of young redwoods forming in this way around the bases of parent trees or stumps (*Figure* 19). These small circles of trees are clonal, meaning they share the same genetic makeup as the parent tree.[3]

It is often assumed that larger circles (~30+ feet in diameter) of old growth redwoods are formed in the same manner, being clones of a parent tree (*Figure* 20). If so, this presents some curious problems.

Figure 18. An ancient redwood cathedral on Esselen land.

Figure 19. Small redwood "fairy ring" formed from clonal growth of adventitious stems around the stump of a parent tree.

Figure 20. Large "fairy rings" of ancient redwoods are not all clones and show no signs of a parent tree.

If these large circles (100+ feet in circumference) were formed in situ around a parent tree, the stump size would be unprecedented. Coast redwoods are big but not that big! However, let's suppose there were a parent tree of enormous size in the past. Considering how resistant redwood is to decay, it might be expected that some remnants of its massive stump would still exist. But never have I observed any trace of a parent tree within these large redwood circles.

Alternatively, perhaps the large circles formed not from an original parent tree, but from an intermediate generation (or generations) of redwoods that expanded outward from the parent tree. However, this explanation is again suspect because there are no remnants of the stumps of these intermediate generation trees either to be found in the large circles.

Finally, it is not uncommon to find single stem genets among the trees in the larger circles. These genets are not clonal and exist as multiple genotypes, meaning they are genetically distinct individuals that either arrived by seed or were planted.[4] This suggests that other factors, unrelated to sprouting from a parent tree, are involved in the creation of the large circles.

By whatever means it was done, and for whatever reasons, many groves of ancient coast redwoods in Big Sur have nevertheless assumed large somewhat circular and semicircular arrangements, neither decreasing nor expanding in size for more than a thousand years, all while under the care of the Esselen.

Fire regimes in redwood forests

Today the ancient redwood groves in Big Sur are heavily overgrown by young redwoods, with new arrivals every year (*Figure* 21). The largest redwoods are showing signs of stress in their upper canopies, and many have lost their tops. One is tempted to think that it was always this way — a few of the young trees eventually grow to replace old dying trees, and the cycle of life continues. But that is not what appears to have happened here.

Figure 21. Redwood grove in Big Sur. This forest was thinned about a decade ago and is already showing signs of being overgrown with young redwoods.

To understand why, one needs a keen sense of size-age relationships of redwoods. And the best way to get this sense is to spend lots of time counting the annual rings of fallen and cut redwoods grown in both open canopy and closed canopy conditions. In doing this, I have found a remarkable age gap of redwoods in the Big Sur groves. The majority of trees have smallish diameters and are less than one hundred years old, and more often less than fifty years old. Most of the large redwoods are in the five hundred to 1,000+ year range.[5] However, there is a missing cohort (or cohorts) between about one hundred and five hundred years. This suggests that the largest redwoods grew most of their lives in more open canopy conditions with limited competition or encroachment from younger redwoods. It seems that maybe you don't need to maintain hundreds of young redwoods to replace five or ten ancient redwoods. While this age interpretation is based on informed observations and ring counts of stumps, it is difficult to assign precise ages to large living redwoods.

When inspecting the stumps and downed logs of large redwoods felled in the past century, semi-concentric tree ring patterns depicting fire events and recovery are sometimes observed. By counting and averaging the number of recovery rings between fire scars, one can determine the mean fire return interval for the stand. Using this technique, the University of California scientists Scott Stephens and Danny Fry studied the fire history of redwoods north of Big Sur, in the Tamien lands of the eastern Santa Cruz Mountains.[6] They found that, between 1615 and 1884, the mean fire return interval in the redwood groves was just twelve years. Another study reports that the mean fire interval for redwoods in the Monterey Bay area during precolonial times ranged from seventeen to 82 years, whereas the estimated natural (lightning-caused) fire interval was 135 years.[7] A review of the precolonial fire history research covering nearly the entire range of redwood forests in California reported that most (eleven out of twelve) studies found fire return intervals of eight to fifty years.[8] This return interval is significantly shorter than would be expected from natural, lightning-caused fires, which is variously estimated at hundreds to even thousands of years in many California redwood forests.[9] In the region around Monterey Bay and Big Sur the estimated natural (lightning-induced) fire return interval in redwood forests is 550 years.[10] Thus, the high frequencies of fire as documented by fire scar analyses in old-growth redwoods can only be attributed to cultural burning by the Esselen and other nearby tribes.

Also revealing is that the redwood forests of the Mendocino region, well north of here, had pre-twentieth-century fire-return intervals ranging between six and twenty years — similar to the Big Sur region. Interestingly, detailed scar analyses showed that the vast majority of these fires were found to occur between late August and late November.[11] However, mean monthly lightning strikes recorded in Yosemite, just 160 miles away, are shown to be highest in June, July, and August, suggesting lightning is not a leading factor in frequency of fall fires in redwoods.[12] Here, then, is further scientific evidence for what TEK practitioners have long known, that the redwood forests are best tended with cultural burns between late summer and late fall. In the case of the Mendocino region, these fire practices would have been maintained by the Pomo.

Native tending of redwoods

The Native Tribes living among the redwoods had many reasons to care for them. The wood and bark of the redwoods were commonly used in the construction of homes, lodges, and roundhouses. Redwood trunks were transformed into seafaring canoes and paddles, and meat platters, finger bowls, and stools were finely constructed from the wood. The dense, rot-resistant wood was used to build acorn granaries that offered protection against pests and mold, and acorns could also be stored in the burned-out trunks of redwoods. Redwood roots also make excellent basketry material, and redwood leaves and pitch can be incorporated into herbal remedies.[13] It is thus reported that the red pitch from the oldest trees was preferred by some Indigenous healers as it possesses greater potency and a longer memory of the forest.[14] There were even ceremonial roles for redwoods, such as the naming rituals of the Rumsen that occurred beneath them.[15] It is also reported that "El Palo Alto" (meaning "The Tall Stick"), a landmark redwood "discovered" by Spanish explorers in 1776, was used as a "council tree" by the Ohlone.[16]

Here in Big Sur many of the giant redwoods bear remnants of large lower branches that could only have formed if, in the not-too-distant past, the trees' lower canopies grew in conditions more exposed to light. Indeed, between some clusters of big redwoods are significant gaps in the forest canopy that contain species-rich remnants of prairie, known in other locations to have the resulted from Indigenous burning.[17] Also found in these gaps are the scattered remains of large tanbark oaks that flourished among the redwoods in the recent past. In effect, many of these areas were previously open-canopy tanbark oak/redwood savanna, rather than the present-day dark, dense redwood forest.

Much like the Esselen, the Yurok People to the north, "regularly burned ground cover in redwood forests to bolster tanoak populations from which they harvested acorns, to maintain forest openings, and to boost populations of useful plant species such as those for medicine or basketmaking."[18]

The above observations make it apparent to me that for a thousand years or more the Esselen employed fire methodically in and around these groves to create a mosaic of ancient tanbark oaks, bay laurels, and coast redwoods. They understood that burning around the trees eliminated competition and fertilized the soils with ash and biochar.

The origins of "goose pens"

Prominent features in many redwood circles are the fire-carved cavities at the base of the trees, colloquially referred to as "goose pens,"[19] which often face inward toward the center of the groves. A careful examination of these burn scars show they are the result of multiple fires occurring over many hundreds of years. Some of the largest trees have been completely hollowed out by fire, creating chambers inside the tree that can hold several people. You would think that burning the entire woody innards of a mighty redwood trunk would undermine the tree's health and stability, yet these fire-scarred trees are among the largest and oldest beings in Big Sur.

It appears to me that the Esselen and other California tribes recognized how the fiery creation of these cavities stimulated the regrowth of bulky scar tissue, resulting in a flaring at the base of the redwoods. In effect, cultural fires were used both to improve the health of the redwoods by reducing competition, and to *structurally engineer* them to have more stable foundations (*Figure* 22) compared to unburned redwoods (*Figure* 23).

Figure 22. A fire-scarred coast redwood trunk showing wound-induced flaring of the base. Note the nearby overcrowding by young trees.

CHAPTER 3

Figure 23. Unburnt coast redwood trunks do not typically flare at the base.

Some insight into understanding the possible cultural significance of these redwood burning practices can be found in the folklore of Native People in the Pacific Northwest. In her book *The*

Mystery of the Fire Trees of Southeast Alaska author Mary Ida Henrikson reports that ancient burned-out western redcedar trees were used to store fire. [20] Her source for this practice was George Eton, a Tsimshian from Metlakatla, Alaska, who still lived a mostly traditional life in the early twentieth century. Properly constructed, the inside of a burned-out, but still living, tree would provide a sturdy shelter from rain where a fire could be tended for months at a time. Many of the burned-out redwoods in California have sheltered cavities that could also have been used to sustain a fire for extended periods.

Henrikson goes on to describe other aspects of "fire trees," including faded ceremonial carvings in the scars. She likewise speculates that certain western redcedar fire trees were strategically located on ridgetops, where their fires served as beacons to guide Native mariners across the wide channels and straits of the Alexander Archipelago. All of this points to a potentially rich Indigenous heritage in the many fire-formed trees of the Pacific Northwest.

Returning to Big Sur, the temple-like appearance of the old-growth redwoods altogether emanate sacredness (*Figure* 18). I have often noticed that, upon entering a large redwood circle, people tend to quiet their voices, as they would when entering a church or other revered space. In most of these groves the insides of several giants are burned out, creating large chambers (*Figure* 24), some even with multiple openings (*Figure* 25). I have visited numerous burned-out redwoods with Little Bear, Esselen elder, and with the late Billy Post, a Rumsen descendent, and both have expressed strong support for the idea that there is a great cultural significance in these fire-carved trees.

The profound point I'm trying to make here is that many of the ancient redwoods, in my opinion, are clustered in cathedral-like groves intentionally — often with circular or semi-circular arrangements and with their fire-scarred cavities facing inward to create hallowed, ceremonial spaces (not "goose pens"). Furthermore, the basal burning of the large redwoods resulted in splayed trunk shapes that better supported the massive ancient trees in these sacred groves, beings that Little Bear refers to as his "Standing Brothers and Sisters." While many of us have felt it, there needs to be a more explicit recognition that the Esselen, along with many other California Tribes, didn't build their temples, they grew them!

I encourage folks, while in the company of ancient redwoods, to mull on the idea that this is all by design, a living reminder of the honor and reverence with which the Ancestors of today's California Tribes regarded the Standing Ones. In my world it is common to spend time in these groves and chambers, touching, smelling, and otherwise sensing the trees, lying on the ground while contemplating the shape and structure of the surrounding spires, and feeling the subtle forces held deep inside an ancient being. In these places I am keenly aware of being enveloped by the earth energy flowing upwards, and by the sun and sky energy flowing simultaneously downward. In time, one begins to see ciphers in the burned-out redwoods. I, for one, can't help but see the tall, erect trunks of the giant redwoods that open at their base into yoni-shaped cavities as fertility symbols, depicting aspects of both the masculine and the feminine in the united form of a mighty forest being (*Figure* 24).

There are thousands of redwood groves in California and Oregon that bear similar structures and arrangements to those in Big Sur. These groves undoubtedly hold rich histories of the land they grow on and the Peoples who cared for them. Whether they be considered temples, memorials, sanctuaries, symbols, living artifacts, or random acts of nature, the giant coast redwoods are the enduring legacies of a caring culture from a time when many more of these ancestor trees stood tall.

Figure 24. Large cavity created by multiple fires inside an ancient coast redwood.

Figure 25. Sacred "Four directions" coast redwood on Esselen land, shown to me by Tom Little Bear Nason.

4

ON THE AGE AND PROVENANCE OF THE MONTEREY CYPRESS

This brief chapter may seem tangential to the topics of cultural fire and California Native tending of trees. But, to be frank, some readers will likely have a fit over what I report in the next chapter on the occurrence of ancient "non-native" Monterey cypress trees growing on Esselen village sites in Big Sur. So let me first address what is known, and what is not known, about the origin of the Monterey cypress forests in this region.

The colonial narrative

The current narrative describing the age and provenance of Monterey cypress is deeply embedded in colonial culture. The story goes like this — today's Monterey cypress trees, which have been extensively planted since the mid-nineteenth century and now grow prolifically along the California coast and elsewhere in the world, have *all* originated from small relic populations occurring at two localities along the shores of Carmel Bay. The Monterey cypress is said to be endemic to the Monterey area and did not previously exist anywhere else in the world outside of these two sites at the time of its reported discovery, which was less than two hundred years ago. The botanist Willis Jepson, in 1910, stated with authority:

> [many] expeditions to the California coast have come and gone, and we now know definitely, after this long period of searching, that the Monterey cypress (*Cupressus macrocarpa*) does not occur at any other locality in California — nor elsewhere in the world.[1]

This story is repeated in every single account and botanical description of Monterey cypress and is essentially Western scientific dogma. To my knowledge this narrative has never been challenged.

It is relevant to know that the Monterey cypress was not always such a rare species. It, along with the Monterey pine, once formed extensive forests in this region, as shown from Tertiary age fossils.[2] During the late Quaternary, the range of Monterey cypress trees remained large, extending from north of San Francisco Bay to Southern California.[3] However, at the end of the Pleistocene ice age, about 12,000 years ago, the California climate shifted from cooler and wetter conditions to warmer and drier ones unfavorable to the Monterey cypress. This resulted in a severe constriction of its former range into the two aforementioned relic stands. In other words, it is believed that Holocene climate change caused the population crash and near extinction of the Monterey cypress.[4]

So, allow me to ask: How, in just two hundred years, did a species on the verge of extinction (with a presumed total population of only about ten thousand individuals[5]) suddenly explode in numbers here and abroad? Most would say it is because they were planted by Western settlers.[6] Which then begs a second question: Why is this species, after being planted, now thriving under the same adverse climate conditions that nearly caused its extinction?

Let me tellingly state that nowhere in this narrative are the Native People said to be a factor in the distribution of the Monterey cypress! There are some accounts that mention the Rumsen People living at times among the cypresses — as indicated by the presence of shell middens, charcoal remains, and grinding stones (which I have personally observed in the Point Lobos groves).[7] Despite clear evidence that there were two "more or less permanent" Rumsen villages nearby, one on either side of Point Lobos, the Western narrative considers these Native People merely as ghosts of the past who lived peacefully near the ocean shore amidst the "untouched wilderness" and "primitive landscape" of these picturesque cypress trees (*Figure* 26).[8]

Figure 26. Relic Monterey cypresses growing among Rumsen shell middens and grinding stones at Point Lobos.

Coincidently or not, during the last few hundred years, while Monterey cypress have greatly expanded their range, there has also been a massive appropriation of California Native lands by Western settlers and a corresponding suppression of cultural burning. This has created a profound shift in land use patterns over the past century. Thus, let me offer a somewhat different narrative, one that includes the wisdom and customs of the Native People.

A Native revisionist narrative

For thousands of years the Rumsen and the Esselen People managed the local ecosystems with fire, using this tool to encourage certain plant and animal communities as well as to limit the growth of other communities. Oak savannas and coastal prairies were favored for their abundance in food resources. Other communities such as chaparral were constrained by the frequent burning. Thus, I believe that the restricted distribution of Monterey cypresses was probably intentional. While cypress trees have traditional uses, which may explain why the Native People did not let them go completely extinct, they do not produce any food or forage.

The Monterey cypress is a fast-growing, coastal species that, at maturity, is known to recover poorly after fire. Although cones of the cypress are somewhat serotinous,[9] neither is the species fire adapted nor is its regeneration fire dependent. Therefore, in a landscape that was managed with fire for millennia, it is not surprising that the Monterey cypress would have a limited range. However, . . .

What was the precolonial range of Monterey cypress?

The answer to this question, despite published claims, is that we don't actually know. I and others find it remarkable that such famed botanists as Archibald Menzies, David Douglas, and Thomas Nuttall who visited Monterey in 1792, 1830, and 1835, respectively — as well as two earlier botanists, Jeanne Baret and André Michaux, from the La Perouse expedition who visited Monterey in 1786 — made no mention of any cypress species.[10] Instead, the first reported encounter by a Western botanist was by Aylmer

Lambert in 1838, who never visited California but somehow obtained and brought a few seeds of this cypress with no name or provenance to the Horticultural Society of London, where they were then propagated.[11]

It wasn't until 1846, less than two centuries ago, that the botanist Karl Theodor Hartweg first described this species in its native habitat and reported it to Western science.[12] We know for certain that in 1846 Hartweg visited only one Monterey cypress population, the Del Monte forest.[13] Otherwise, there is no record of when the Point Lobos cypresses were delineated, nor by whom.[14] Most modern-day discussions of the origin of the Monterey cypress reference Griffin and Crutchfield or MacBride,[15] who both mention groves of cypress trees on either side of Carmel Bay. Yet who among the eighteenth and nineteenth century botanists actually searched the rugged coastlines and canyons of Big Sur and other nearby areas for the presence of this species? Instead, records of the Monterey area explorations, as noted above, indicate that the botanists prior to 1846 were severely limited in their forays. They thus returned with no records of the Monterey cypress, despite the easily accessible "relic" cypress populations along the shores of Carmel Bay. This makes me suspect that there were other such relic stands "undiscovered" on Esselen lands as well.

It seems relevant to note, too, that in the Esselen language there is a word "*tsumir*," meaning "cypress,"[16] likely a reference to the Monterey cypress, which (presumably) did not grow on Esselen lands. It is possible that *tsumir* alludes, instead, to a relative of the cypress, the California juniper, which has a narrow range in the Arroyo Seco area of Esselen territory. But in the next chapter I will describe Monterey cypress trees that, based on their large basal diameters (greater than five feet), are apparently well over 250 years old growing on Esselen village sites not far from these "relic" groves. This conclusion, however, is not possible according to the current narrative, which states that all Monterey cypress trees in Big Sur were planted by colonists and are, thus, "non-native."[17]

CHAPTER 4

. . .

Missing age data

One way to sort out this discrepancy is to examine age structure data of Monterey cypress populations to determine the timing of their establishment. Unfortunately, critical data on the ages of the largest cypress trees are completely lacking in the scientific literature, save for a single citation by Jepson of a definitive age for one relic Monterey cypress tree, a respectable 284 years old.[18] This I would refer to as an ancestor tree, as it dates from the time when the Rumsen were the sole human occupants of Point Lobos. Sadly, no information was given on the size or shape of this tree. Jepson also reported one other tree age taken from a younger cypress that was cut after falling over a road. It was found to be 98 years old with a diameter of two feet.[19]

Still there is a way we can estimate the ages of the large cypress trees, by examining age and size structure data in younger populations of Monterey cypresses planted in the San Francisco parks during the 1880s.[20] From this data, making the standard assumption that there is a gradual decline in diameter at breast height (dbh) with age, I conservatively estimate that a Monterey cypress with a five-foot dbh would, under *ideal* growth conditions, be at least 250 years old and perhaps much older. This age estimate is corroborated by a simple extrapolation of Jepson's age of 98 years for a two-foot diameter cypress. All of these findings suggest, to me, that many of the cypress trees growing on Esselen village sites in and around Big Sur likely established long before the arrival of colonists.

Of course, we won't know for sure until someone gathers age and size structure data on the local Monterey cypresses, both from their relic stands and from other nearby cypress giants, of which there are many. I'm sure an inquisitive student of forest ecology would find this to be an exciting field project in Big Sur!

5
ANCIENT TREES AND SHELL MIDDENS

Scattered along the coast of Big Sur are hundreds of shell middens, most of which are recognized by the Esselen Tribe as former village sites. My Esselen friends are often called upon to inspect these sites and occasionally discover human burials in the middens, revealing how hallowed they were, and still are, to the Esselen tribe.

Upon many of these middens also grow old Monterey cypress, coast live oak, and coast redwood trees (*Figure* 27).

Figure 27. Ancient coast redwood growing atop a shell midden on Esselen land.

CHAPTER 5

. . .

Of ravens and raptors

There is a backstory to how I began to sense an affinity of ancient trees with shell middens. During my first summer working as a high-lead logger in southeast Alaska in 1978, I encountered a young fellow named John Muir, an itinerant deckhand who had the same scruffy appearance and unique character that I imagined the nineteenth-century John Muir would have had. He was camped beneath a giant Sitka spruce, along the banks of the Indian River outside of Sitka, Alaska. One day, while visiting him at his camp site, I inquired about the hundreds of fragments of seashells scattered on the ground and commented that I had seen shell fragments around other large trees in the area. He said that this was due to the bald eagles, ravens, and gulls harvesting shellfish. The birds first dropped them onto rocks to smash the shells, then carried the remains to the high treetops, where they would feed and drop the shells. This sounded plausible at the time, and I did not think about it again until… nine years later when I was in the late stages of my doctoral research.[1]

At that time, I found myself on Kruzof Island, across the sound from Sitka, completing the last of dozens of soil profile analyses which started near the summit of Mt. Edgecumbe and extended downslope along a nine-mile transect that culminated at the Pacific coast. While digging a soil pit in a grove of enormous Sitka spruce and western hemlock trees, the largest by far of any of the 12,000 trees I had surveyed, I encountered something my scientific training had not prepared me for — a layer of seashell fragments about eight inches below the soil surface. I distinctly remember my first thought, "Is this evidence of an old tsunami?" But the stratigraphy of this soil profile and of soil profiles nearer the coast showed no signs of such a catastrophic event. There was only this one site, with a thin layer of crushed shells around several giant trees.

So where did the layer of seashells come from? It was then I suddenly recalled the shells around the big trees along the Indian River and had an inkling that something greater was at play here than could be attributed to the leftovers of ravens and raptors. Later, I learned that the shell layer I had found is what some would call a "sheet midden," deposited hundreds of years ago by the Tlinkit Indians. This was, truthfully, another occasion where John Muir had misinformed me about the importance of the Native Peoples in the creation of "the wilderness."

What are shell middens?

Shell middens are ancient accumulations of saltwater and freshwater mollusks, fish and other animal bones, stone flakes, pottery, and, sometimes, human remains. They are often intermingled with layers of dark organic-rich ash and charcoal. Shell middens are found on every continent, except Antarctica, mainly in coastal settings, although sizable middens consisting of freshwater mollusks can be found hundreds of miles inland. Some middens are exceptionally large, such as the mega-middens of South Africa, which are comprised of hundreds of thousands of cubic feet of shell material.[2] Another well-known example, Turtle Mound in Florida, contains nearly one million cubic feet of shells and had an estimated height (before modern excavation) of 75 feet. Turtle Mound was so tall that early mariners used it as a landmark for coastal navigation.[3]

Archeologists have typically referred to shell middens as "garbage" or "refuse" piles, accumulations of the discarded remains of daily life. Danish archeologists in the 1800s coined the term "midden" from the Danish *mødding* or "heap of dung." The original reference was to ancient shellmounds they discovered in Denmark, which they called "kitchen middens" arising from the leftovers of ancestral human occupations.

My Esselen friends, as well as some archeologists, however, say middens are not sites of waste disposal, rather they are former village sites and/or culturally significant centers of ceremonies, feasts, and burials.[4] Some Indigenous Californians have pointed out to me that they have no words in their language for "garbage" or "refuse." In this regards the term "midden" is contextually inaccurate, as middens were never merely rubbish piles, as will be discussed below. Shell middens are also referred to in the archeological literature as shell-heaps, shell-matrix sites, sheet middens, and shellmounds, although not all middens are mound-like.

I have pointed out in a previous work a number of enigmas about shell middens.[5] In his 1906 report on the excavation of the Emeryville shellmound in Berkeley CA, Nels Nelson noted that the structure of the mound "presents some curious problems," the most remarkable of which is its great size.[6] Originally reported to be more than sixty feet high and 350 feet in diameter, the Emeryville shellmound was a significant feature of the coastal landscape, and was linked to several adjacent shellmounds by extensive low-lying sheet midden deposits.[7] Interestingly, Nelsen also found evidence that the mound had been periodically excavated and reworked during its formation, and that much of the substructure of the mound was likely "not in the place of its original deposition."[8] Knowing the high fertility of these nutrient-rich deposits, the first question I would ask is: Were middens deposits used by the Ohlone People to amend the soils of their gardens and crops?

Shell middens in Native cultures

In 2006, considering the evidence cited above, I hypothesized that, "the many types of refuse mounds and middens, including shell middens, bone middens, and rock middens originate not from the gradual accumulation the waste products of daily living, but rather from the intentional stockpiling of gathered or recycled lime-rich materials for use as mineral fertilizers."[9]

Since then, I have become better aware of the broader cultural significances of shellmounds and middens to Indigenous Peoples. Human remains, many in burial positions, occur in shell middens in California and elsewhere. Shell middens were also used for gatherings and ceremonies. In his detailed work on ring middens of the southeastern U.S., Michael Russo concluded, "other than simply trash piles, middens can serve multiple purposes such as monument construction, symbolic signification, and territorial and social status marking."[10]

Other reports of shellmounds in Florida reveal that the sites were used for summer solstice gatherings and feasts.[11] The Garden Patch shellmound site along the Gulf coast of Florida is similarly described as "a civic-ceremonial center."[12] If shellmound sites were used for such purposes, would it not make sense that these gathering sites supported some large trees for shading participants?

In a 1903 account from the Jesup North Pacific Expedition of the lower Fraser River in British Columbia, H.I. Smith described Douglas fir trees more than seven feet in diameter growing upon shell middens. On the size of these middens, he remarked: "The typical shell-heap is several hundred yards in length, about thirty yards in width, and three to four feet in height. Others are miles in length, and some reach a height of over nine feet."[13]

The close association of culturally modified trees (CMTs) with shell middens is another indication of the societal significance of these sites. Archeologists have identified numerous culturally modified southern beech trees growing at ancient Kawésqar shell midden sites in western Patagonia.[14] In 2016, scientists from British Columbia published an analysis of western redcedar trees and found many CMTs on midden sites. They concluded, "Western redcedar trees growing on the middens were found to be taller, have higher wood calcium, greater radial growth and exhibit less top die-back [compared to adjacent non-midden sites]." They further stated that this was an example of "long-term intertidal resource use [shell middens] enhancing forest productivity and we expect this pattern to occur at archaeological sites along coastlines globally."[15]

This assertion should be foremost on the mind of any archeologist studying shell middens. Are or were ancient trees conspicuous features of other shellmounds and middens throughout the world? (*Figure* 28).

Figure 28. Ancient tree (southern live oak?) growing on a shell midden in Coden, Alabama circa 1940 (photo downloaded from www.gulfsouthpastrecovery.com/tag/shell-middens/).

Several studies report that shell middens not only have a positive effect on tree health, but they are also centers of both plant domestication and diversification. In his book *Cultivated Landscapes of Native North America*, William Doolittle describes the "dump-heap hypothesis," proffered by Edgar Anderson in 1952.[16] This proposed that the highly fertile midden soils near habitation sites served as ideal places for plant domestication.[17] Doolittle also provides a detailed account of ridged Native American agricultural fields in Wisconsin, and observes how "The ridged fields were also found to be covered in midden debris consisting of charcoal, ash, fish bones and shell, suggesting that composting or fertilizing may have been employed."[18]

The idea that midden soils were centers of plant cultivation has been recently investigated in the Amazon basin, where highly fertile dark earth soils (i.e., middens) were found to have been intentionally created in regions of plant domestication.[19]

A study by W. McAvoy and J. Harrison on the plant biodiversity of shell midden sites in Maryland reported:

> Fourteen Native American shell-middens were discovered on the Delmarva Peninsula in Kent, Queen Anne's and Dorchester Counties, Maryland. Occupying these shell-middens is a unique and globally rare plant community that supports 202 native species and varieties of vascular plants, including 87 that are rare or uncommon on the Peninsula and 21 that are new additions to the flora of the Delmarva.[20]

In the same region, a study by S. Cook-Patton et al. looking at the soils and plants of ten shellmounds along the Chesapeake Bay, also concluded that the ancient shell middens exhibited higher levels of soil nutrients, more plant cover, and greater species richness than adjacent non-midden sites.[21]

Working in Baja California, S. Vanderplank et al. also found significantly higher plant diversity at two shell midden sites compared to adjacent sites.[22] A study by J. Fischer et al. examined the biodiversity of ten "habitation sites with extensive shell middens" in coastal British Columbia, and likewise reported a significant enhancement of soil fertility and biodiversity compared to adjacent "non-midden" sites. They also found that these habitation sites strongly favored culturally important plant species.[23] Conversely, in a study of five habitation sites in coastal British Columbia, K. Schang et al. reported slightly less species diversity at habitation sites vs. control sites. They did, however, find that, for all tree species, maximum height was significantly greater on habitation sites compared to control sites.[24]

Regarding soil properties of middens, scientists analyzing shell middens in the state of Georgia in the U.S. have reported that midden soils, with a mean pH of 6.7, had calcium concentrations ten-fold higher than adjacent acidic soils [25] Along these same lines, Trant et al. stated that enhanced forest productivity in British Columbia was associated with higher pH and greater availability of soil phosphorous and calcium at shell midden sites. As they concluded, "With a deep time perspective from 13,000 years of repeated occupation of the study area, it is clear that coastal First Nations people have developed practices that enhanced nutrient-limited ecosystems, making the environment that supported them even more productive."[26]

The above evidence, and more,[27] points to the profound influence of shell middens on tree health, plant diversity, and soil fertility. These properties of shell middens could not have gone unnoticed by their creators. Beyond this, however, I believe the simple fact that so many shell middens were destroyed by settlers, who used the midden materials to fertilize their fields, makes it obvious that shell midden soils have long been recognized as a rich source of biochar and nutrients in agriculture.

Returning to Big Sur, at several coastal Esselen village sites, ancient trees may still be found firmly established on old shell middens (*Figures* 29 and 30). Indeed, four of the largest Monterey cypress trees that I have seen in Big Sur grow atop shell middens. And the biggest are very similar in size and shape to the largest cypresses at the nearby Point Lobos "relic" groves, which are also growing on shell middens. Interestingly, these do not take the typical tall cypress form, with a single main trunk, but are shorter and more heavily branched, with broad spreading canopies. They, like the oaks, thus appear to have been culturally modified and tended by the Esselen people, only not for the purpose of food or forage production. The Monterey cypress generate large cones rich in pitch, which can be used as an adhesive and sealant. Cypress leaves and bark also have antiseptic properties and are widely used in traditional folk medicine.[28] However, the key reasons for tending Monterey cypress are that they afford protection from strong coastal winds and offer broad, shady gathering places out of the summer heat.

Figure 29. Large Monterey cypress growing on an Esselen shell midden/village site.

CHAPTER 5

Figure 30. Esselen shell midden soils beneath the Monterey cypress shown above (*Figure* 28).

An archeological study in the Big Sur region found that while shell middens tend to be clustered along the coast, they also occur well inland. Some of the middens are located twelve to eighteen miles inland and up to 2,800 feet in elevation. Radiocarbon analyses indicate that these middens date to about 6400 years BP (before present).[29] The radiocarbon data reveal, too, that the materials in the middens are commonly out of stratigraphic order, meaning that older materials are sometimes found overlying materials significantly younger in age. Here, again, is evidence that middens were disturbed or reworked, suggesting that these deposits were being dug up for some purpose.[30]

Tom Little Bear Nason states he has found Esselen shell middens thirty to forty miles from the coast, as well as near the 6,000-foot summit of Junipero Serra Peak (*Pimkola'm*), the highest peak in the Santa Lucia Mountains. He also describes a twenty-foot-deep shell midden that occurs in Cachagua, about twelve miles from the coast in northern Esselen territory. Tribal archeologists, led by Jana Nason of the Esselen Tribe's Department of Cultural Resources, participate in all excavation activities at these shell midden sites and closely monitor the material removed for artifacts and evidence of human remains. The sanctity of shell middens in the Esselen culture is echoed in the beliefs of many other Indigenous Tribes, who view shell mounds and middens on their lands as sacred sites.

The relevance of shell middens and their positive effects on soil fertility and tree health will also become clearer in later chapters of this book, in which I will describe the use and efficacy of seashells, ash, and biochar amendments in the practice of fire mimicry.

6

THE CRYPTIC ECOLOGY OF MOSSES AND LICHENS

In nearly every forest that is exhibiting symptoms of decline there are a group of inconspicuous organisms that are rarely studied or even mentioned in the pathological or ecological literature. I refer specifically to the mosses and lichens, which are considered cryptogams (spore-reproducing organisms). Most ecologists apparently deem these organisms relatively insignificant, given that they are generally not recorded in ecological surveys, or are classed as litter or soil. They are typically treated as organisms that occupy niches which other species ignore — ecological gap-fillers if you will, with marginal influence on the overall function of the ecosystem.

While often found growing together, mosses and lichens have distinctly different evolutionary origins. Mosses are bryophytes, nonvascular plants that are closely related to the earliest known land plants, the liverworts.[1] Most mosses have leaves, stems, and root-like structures called rhizoids, which serve to anchor the plant to the substrate, and reproduce both asexually (e.g., layering) and sexually. Sexual reproduction of mosses occurs via spores that are propagated in stalk-like structures called sporophytes. But this mode of reproduction can only happen when there is sufficient moisture in the environment (*Figure* 31).

Figure 31. *Cynclidium* spp. mosses.

Lichens, on the other hand, are a curious combination of various algal and fungal species growing in close symbiosis (*Figure* 32). Although often classed as members of the plant kingdom, lichens are comprised mainly of fungi and only 5 percent or less of algae. They have photosynthetic thalli that take on a wide range of shapes and colors. They are typically leaf-like (foliose), branch-like (fruticose), or crust-like (crustose), with rhizines that anchor the plant to the substrate. Lichens can reproduce sexually, via spore-filled apothecia, or vegetatively, whereby both the fungal and algal components of the organism are dispersed together in propagule-bearing structures such as soralia, soredia, and isidia.

CHAPTER 6

Figure 32. Foliose lichen (*Flavoparmelia* sp.) growing in a mat of mosses (*Grimmia* sp.) on the bark of a coast live oak.

Both mosses and lichens are relatively small, evergreen, and perennial, and can grow in nearly any habitat, warm or cold, wet or dry. Mosses and lichens are either ground-dwelling (growing on soil, rock, or fallen logs) or epiphytic (growing on the trunks and branches of other living plants). They are considered nonvascular, meaning most do not have a system of vessels to conduct water and nutrients. Instead, they transport water and nutrients via a process of controlled diffusion through individual cell walls and membranes.

While mosses and lichens obtain their energy from photosynthesis, they still require a suite of essential nutrients that must be absorbed from their environment. Some nutrients arrive via rain, aerosol particles, and dust, while others are absorbed from the substrate upon which they grow. To be clear, *all* mosses and lichens take in nutrients from their surroundings, whether that is air, water, rock, soil, wood, or bark.

Many species of mosses and lichens can tolerate long periods of freezing and desiccation but will quickly revive when temperatures warm and/or when moisture arrives. In cold environments this gives them a competitive edge over neighboring vascular plants, especially in the spring when mosses and lichens can take immediate advantage of the spring sun and nutrient-rich waters from the melting snows (even starting to grow beneath the snow), while nearby plants with their roots still frozen remain dormant. Mosses, too, are known to be effective insulators, thus lowering subsurface temperatures and lengthening the period of frozen soil conditions.[2]

There are various beneficial roles that mosses and lichens serve in forest ecosystems. Their spongelike bodies hold an abundance of moisture that, upon evaporation, acts to cool the forest atmosphere. They also provide habitat for arboreal fauna such as salamanders and tree frogs, as well as offering nesting materials for birds and forage for deer. Mats of feathermosses on the forest floor provide stable, moist substrates for the germination of some, but not all, tree seedlings. Many moss and lichen species also support various kinds of cyanobacteria that fix nitrogen, thus providing an additional source of this essential nutrient to the forest ecosystem.

There are, however, properties of mosses and lichens that can pose problems for tree health and soil fertility. In their attempts to grow, lichens and mosses exude acidic compounds that break down substrates and free essential nutrients. Given that mosses and lichens are known for their ability to decompose rocks to form soils, one must wonder about the damage they may be doing to the bark and roots of living trees.

What follows are germane observations and results that I and others have obtained by carefully examining the roles of these cryptic organisms in shaping the structure and development of the ecosystems where they are abundant. As will soon become evident, the subtle but coherent forces of mosses and lichens on the health of trees, the fertility of soils, and the succession of ecosystems are not to be underestimated.

Lichen effects on trees

The Native People seem to have been well aware of the ecology of lichen epiphytes. Describing Native use of fire in the forests along the present-day Montana-Idaho border, an explorer's journal from July 1860 states:

> In returning, the Indian set fire to the woods himself, and informed us that he did it with the view to destroy a certain kind of long moss [not a moss but an epiphytic lichen] which is a parasite to the pine trees in this region, and which the deer feed on in the winter season. By burning this moss [lichen] the deer are obliged to descend into the valleys for food, and thus (the Indians) have a chance to kill them.[3]

This account suggests that Native Americans recognized two fundamental ecological processes: 1) that lichens are harmful to trees, and 2) that fire reduces lichen cover.

The earliest reports I can find on the harmful effects of lichens in the Western scientific literature come from D.C. Peattie who, in 1950, wrote about the long trailing beards of lichens that typically clothe old specimens of Monterey cypress. As he writes in his book *The Natural History of Western Trees*: "The *Ramalina* [*Ramalina reticulata*] is indeed an enemy of the Cypresses, for it smothers its foliage and, always gaining headway, eventually kills the tree without actually parasitizing it, but by a sort of suffocation."[4] And he goes on to morbidly describe a similar situation with nearby Monterey pines, also draped in *R. reticulata* lichens: "Though it is not a true parasite, but merely a perching plant, the lichen harms the Pines mechanically by shutting out the light and blanketing the leaves, so that boughs are smothered and starved to death, and sometimes whole trees may die from the high cost of playing host to this dependent."[5] Coast live oaks in this same region appear to be similarly affected (*Figure* 33).

Figure 33. Dying ancestor coast live oak "smothered" in *Ramalina reticulata* and *Usnea* sp. lichens.

More recent studies have examined the chemical and ecological effects of epiphytic lichens such as *Evernia* spp. and *Usnea* spp. and found that they and many other lichen species produce acids, namely evernic and usnic acids,[6] which are known to have detrimental effects on trees, especially oaks. Findings specifically indicate that the lichen *Evernia prunastri* releases evernic acid into the xylem of Spanish oaks which inhibits both respiration and the appearance of foliar buds, as well as slowing leaf formation.[7] Evernic acid is also tied to chlorophyll depletion and decreased photosynthetic activity in the leaves of holm oaks.[8] Other researchers report that lichen penetration through the bark produces clear symptoms of chlorosis and aging of leaves by injecting metabolic inhibitors, often inducing early abscission (i.e., leaf fall).[9] With regard to usnic acid, studies have shown it, too, is a growth inhibitor that negatively affects root length, shoot length, and the root-to-shoot ratio of Scots pine and Norway spruce seedlings.[10]

Given that lichens are sources of acidic compounds known to have direct impacts on trees, I wondered whether there might be other, indirect impacts of lichens in forests. Specifically, I asked: Could acidification of canopy throughfall, caused by lichen substances absorbed by rainwater, enhance the leaching of nutrient cations from the forest soils below? To me this seemed quite plausible, so I performed a cursory experiment.

In the late summer of 2004, while visiting my dear friend Jeff "Willy" Wilson who lives along the Rogue River in Oregon, I tried a simple test on the epiphytic lichens growing on his sick Oregon white oak trees. Using a high-quality pH meter with buffered standards, I first determined the pH of the distilled water used as a solution. It measured 5.6, typical also of atmospheric precipitation. I then added a handful of lichens (*Usnea* sp.) gathered from the branches of Willy's white oaks and remeasured the pH of the solution after a few minutes. This time the pH read 3.6, about two orders of magnitude more acidic than the original solution. From this I concluded that lichens likely do release acidic compounds during precipitation events.

I have searched the lichen literature and reached out to the professional lichenology community, but have found no works that demonstrate, as is often claimed, a benign effect of lichens on trees. Considering the reports described above on the detrimental impacts of lichens, both direct and indirect, it seems fair to say that *excessive* lichen cover can and does harm trees.

Moss effects on trees

Next, let me share some important, but seemingly overlooked, findings on the effects of mosses on trees. A recent, detailed review by the bryophyte ecologist Janice Glime on the roles of mosses in forest health concludes that there are both positive and negative interactions, depending on the tree species, the ecosystem, and the moss species.[11] Overall, however, she describes mosses as "ecosystem engineers," stating that "[Mosses] through their activities in moisture retention, nutrient sequestering, and temperature modification make it possible to sustain (or deprive) mature forests and to promote or exclude seed germination and seedling development."[12]

For instance, in the northeast U.S., eastern hemlock trees frequently regenerate successfully in moss mats on soil, rocks, and fallen trees.[13] However, in the Pacific Northwest, common feathermosses (e.g., *Hylocomnium splendens*, *Pleurozium schreberi*) growing on soils would seem to provide similar moist, suitable beds for Sitka spruce and western hemlock seedling germination and growth, but their seedlings rarely survive.[14] It has also been observed that *P. schreberi* and its fungal associates exert a "powerful control" over regeneration in Scots pine by inhibiting seedling establishment and interfering with the availability of nutrients to seedlings.[15] Furthermore, a study of six different moss species on the regeneration of ten vascular plant species, including two tree species (Scots pine and downy birch), found that, in every case, the mosses suppressed the regeneration of the plant species tested, and that this effect was likely due to reduced soil temperatures created by the mosses.[16]

Ground-dwelling mosses, especially peat mosses (e.g., *Sphagnum* spp.) growing around trees, have been found to acidify the soils and damage fine roots. My and others' field studies in forests of Alaska, New York, Venezuela, and Colorado, which have examined the abundance of tree roots under moss-covered vs. mostly moss-free sites, show that mossy sites, especially those with *Sphagnum*, have significantly less live fine root biomass and significantly more dead fine root biomass than non- or low-moss sites.[17] I have also engaged in controlled-environment greenhouse studies at NCAR on red spruce saplings, which show a detrimental effect of mosses around experimental trees, while control trees growing in soils without mosses appear healthy (*Figure* 34).

Figure 34. NCAR phytotron experiment examining the effects of mosses (mainly *Ceratodon purpureus*) on the growth and health of red spruce saplings. This experiment was assisted by Carl Etsitty, Shaan Bliss, and Jack Wainwright.

Furthermore, data show there is a substantial decrease in the ring widths of trees growing near species of *Sphagnum* mosses (*Figure* 35). Working on Mt. Edgecumbe, Alaska, I found that the growth of all tree species in the study area (Sitka spruce, western hemlock, mountain hemlock, Alaska yellow cypress, and lodgepole pine) was strongly inhibited by moss covered soils, as indicated by the annual average ring widths of trees growing in low *Sphagnum* cover soils (less than 10 percent) being nearly twice the width of those growing in high *Sphagnum* cover soils (70 to 90 percent).[18] I have also reported age and size structure data from black spruce and eastern larch forests in northern Quebec showing, again, that the annual average radial growth rates for both these species were only half in the *Sphagnum*-covered soils compared to the *Sphagnum*-free soils.[19] Of course, I only confirm here what George Rigg reported over a century ago, that there is a marked decrease in the growth of trees surrounded by peat mosses.[20]

The mechanisms by which mosses contribute to fine root mortality and decreased growth of trees are not entirely clear but are likely linked to several factors: 1) acidic leaching of alkaline mineral nutrients from the soils, 2) allelopathic chemicals released by the mosses,[21] and/or 3) the mobilization of iron and aluminum ions that can reach toxic concentrations in soils beneath moss mats.[22]

Figure 35. *Sphagnum fuscum* moss with sporophytes.

Epiphytic mosses, too, can impact a forest in several ways. By catching precipitation and then slowly releasing it to the atmosphere, they can cool canopy environments. As stated earlier, the evaporative cooling by mosses may often be beneficial, especially in hot, humid forests. However, in warm and seasonally dry forests such as the oak forests of California, the precipitation capture and evaporation by mosses (and lichens) may lead to a net loss of moisture from the ecosystem. While little is known of moss-mediated evaporation in broadleaf oak forests, studies in boreal forests indicate that under conditions of heavy moss cover, up to one-third of the moisture in the forest can be lost to evaporation by mosses.[23] In California, the added moisture loss from moss buildup seems to be occurring at the same time that the rampant growth of young trees is further reducing soil moisture. This may even lead to many formerly perennial streams becoming ephemeral in the dry season. Under a cultural fire regime, neither the young trees nor the mosses would be growing in excess, and the streams would likely run more regularly.

The flow of nutrients into a forest ecosystem can also be

strongly affected by mosses. Epiphytic mosses obtain most of their nutrients from dust and precipitation in the atmosphere. When captured by mosses these nutrients enter the forest nutrient cycle in relatively stable forms, tightly bonded with moss tissue that is very slow to decompose and release the nutrients, especially calcium and phosphorus.[24] In some cases, mosses may not completely decompose for a decade or more. The question thus remains: Does this slow release of nutrients ultimately benefit or hinder the health of native forests in California?

Epiphytic mosses are also known to produce significant amounts of organic acids (e.g., malic, oxalic, acetic, and formic acids), which may have detrimental effects on trees (*Figure 36*). While this has not been well investigated, degraded bark on living trees is commonly found beneath mats of epiphytic mosses. In some places, the bark underlying the mosses may be so decomposed that earthworms can be found dwelling beneath it several feet above the ground. In forests where fire has been excluded, mosses tend to be found more frequently on old trees compared to young trees.[25] This is likely due to the slow-growing moss species having more time to colonize and spread on the older trees.[26]

Figure 36. A dense cover of mosses and lichens growing on the trunk of an ancient culturally modified coast live oak. Note the deep cracks in the bark surrounded by thick mats of moss.

Mosses and lichens in succession

My research into the ecology of mosses and lichens began in the arctic tundra on the North Slope of Alaska in the early 1980s, and then expanded into the temperate rainforests of Southeast Alaska and the subarctic peatlands of northern Canada.[27] At all these locations mosses and lichens were extremely abundant, making up a significant portion of the species diversity and green biomass.[28] Furthermore, these regions are quite humid and have been generally free from large-scale disturbances, especially fire, for hundreds if not thousands of years.

In the arena of ecosystem development (i.e., succession) occurring on newly-exposed landscapes (e.g., glacial) that have remained relatively undisturbed for millennia, mosses and lichens start as early arrivals on barren lands thanks to their plentiful, free-ranging spores. But as other key characters, the grasses, the forbs, the shrubs, and the trees, come and fade in this seemingly never-ending life drama that can carry on for centuries, even millennia, the mosses and lichens patiently and steadily expand their roles to eventually become the principal players in the climax scenes of this ecological spectacle — the formation of peatlands. In these ancient ecosystems the mosses, especially *Sphagnum* spp., dominate the biomass, profoundly affect soil chemistry, alter hydrological flows, stunt the growth of trees, and cool the soils and atmosphere (*Figure 37*).[29]

Figure 37. Ancient (~8,000-year-old) peat bog with scattered stunted and dying trees in southeast Alaska.

The idea that peatlands, particularly peat bogs, are climax ecosystems has been debated in the ecological literature for more than a hundred years. Whether or not they are considered "climax" (sensu Clements[30]), "anticlimax," or the result of "retrogressive succession," moss-dominated peat bogs exhibit all the traits predicted of climax communities, including creation of a slow and steady metabolism, alteration and stabilization of the environment, immense longevity, high stores of biomass, and resistance to disturbance.[31]

To me, peatlands are much more than senior-age ecosystems. They represent major organs of the living earth (e.g., Gaia, the Great Mother) that cool the planet by decreasing surface albedo, promoting cloud formation, enhancing permafrost growth, and reducing greenhouse gases in the atmosphere through their ability to store vast amounts of carbon in peat soils.[32] In fact, presently there is more biomass in peatlands than in all the forests and other ecosystems of the world, including oceans, combined!

The lessons I have learned from my investigations into ecological succession are that, just like cells and organisms, ecosystems also undergo developmental life-cycle stages involving birth, infancy, adolescence, maturity, senescence, and regeneration. In ecosystems with high disturbance regimes, succession drives the changes that occur between disturbances. Here in Big Sur, in places that have not burned for many decades, rapid successional shifts in stand structure and species composition are taking place. Succession in unburnt oak savannas is manifest in the proliferation of woody shrubs and young oak trees. Succession in unburnt oak forests is resulting in the replacement of mature oaks by bay laurels and conifers. And succession in unburnt native prairies is evident in the rapid invasion of chapparal and neighboring forest species.

A characteristic feature of succession in undisturbed forests is the gradual buildup of mosses and lichens on the trees along with the acidification of soils.[33] As discussed above, mosses and lichens play critical roles in the shift from forest to peatland. However, the rapid buildup of mosses and lichens in forests that, at least in the near term, are not destined for peatland formation appears to be quite problematic in places. The next chapter elaborates on these and other ramifications of unhindered successional changes due to fire suppression in the native forests of California.

7

A CURRENT UNDERSTANDING OF FOREST HEALTH AND FIRE ECOLOGY

Much of the following information I have touched upon previously, but I wish to distill in this chapter the points that are central to the science of fire mimicry. It is clear to me that fire suppression has led to a buildup of woody understory and an increase in tree density, especially of young trees, in the native forests of California. Under these conditions the native species of oaks, madrones, pines, and redwoods eventually become stressed by the excess competition for moisture, nutrients, and light — and many die. It is known that overcrowding makes trees less tolerant of drought, pests, and wildfires.[1] Therefore, not only is the fire hazard increased by the presence of all the young trees, but it is also enhanced by the addition of dead trees and branches to the canopy and forest floor.

Forest decline

Forest decline, also known as forest dieback, is a global phenomenon that has been attributed to various causes, including climate change, drought, pests, and pathogens, which are not mutually exclusive. The widespread mortality of oaks in California is one such case of forest decline, and evidence suggests that all the above factors are involved. Furthermore, in Big Sur, as in many other places, tree decline tends to affect mature and old-growth stands of oaks more than recently burned, young stands, indicating that successional status is also a factor in oak decline.

Declining forests are often observed to be closely associated with ecosystem age and soil acidification,[2] and typically show a greater buildup of epiphytic lichens in the tree canopies compared to nearby healthy forests.[3] As mentioned previously, acidification of aging soils is a general property of succession and is attributable to an excessive buildup of organic matter, which releases acidic compounds into the soil.[4] Along with acidification, calcium depletion is also observed in areas of forest decline, especially in forests affected by acid rain.[5]

This in not to say that the buildup of organic matter is detrimental to the fertility of soils. In fact, it is a necessary condition for a healthy forest — up to a point. However, once organic debris and matter is allowed to accumulate, rather than being occasionally removed by fire, soil acidification and declining tree health are inevitable.

Any forest experiencing elevated decline and death becomes highly susceptible to catastrophic fires, often destroying the entire structure of the forest. The decline of forests and the extreme wildfire situation here in California have been attributed to a multitude of proximal causes.[6] However, in my opinion the ultimate cause of these destructive wildfires is that the trees and forests in California, which were previously tended by the Native People with cultural fires, are, for the most part, no longer being tended.[7]

Catastrophic wildfire vs. cultural fire

It is estimated that, prior to widespread Western settlement, about 4.45 million acres of California's wildlands burned annually, mostly in cultural fires.[8] For perspective, from 2000 to 2019, the average area burned annually in California was 754,768 acres, or just 17 percent of the pre-settlement annual average. Then, in 2020 and 2021, much more extensive wildfires began to occur, with 4.40 million acres burning in 2020 and 2.57 million acres in 2021.[9] Of course, simply comparing pre-settlement vs. modern burnt acreage obscures a more relevant effect, which is the destructiveness of the fires. Modern fires are burning with severities far greater than previously seen.

For instance, since 2015, extreme fires have killed thousands of native giant sequoias, totaling about one-fifth of the remaining population of these signature trees.[10] And, as Nate Stephenson, an emeritus scientist in forest ecology at the U.S. Geological Survey and an expert on giant sequoias, observes, "What is new and shocking is these large areas, one hundred acres or more, where every single sequoia is killed… [and] there is no evidence anything like that has happened in the past one thousand years, probably many thousands of years."[11]

I am now convinced that the ferocity of modern fires has its origins in colonialism and the corresponding cessation of millennia-long cultural burning practices — an outcome that was enabled by the often-violent actions of European settlers and the unjust politics of their governments. Since then, rapid and profound changes have occurred in forest structure and composition to an extent not seen since the arrival of humans in California more than 12,000 years ago. As noted previously, in some places oak trees are now reproducing faster than the ecosystem can accommodate, creating dense stands of young trees that are encroaching on and stressing the older, mature ones. In other places the oaks are being replaced by later successional species, such as bay laurel, coast redwood, Monterey pine, and Douglas fir. Woody chaparral species have also, over the course of a few decades, begun to invade and replace coastal prairies, which are probably the most endangered native ecosystems in California due to their former heavy reliance on cultural burning. Ancient redwood forests are likewise becoming populated with far too many young redwoods, and the old trees are, thus, suffering. This is starkly evident at the tops of the tallest redwoods, which are displaying dieback due to water and nutrient stress from so many smaller trees growing nearby. Some old-growth redwood trees are also dying entirely, presumably from over-competition (*Figure* 38).

Overcrowding and top dieback is also seen among the giant sequoias. Ancient giant sequoia trees which grew in open conditions for thousands of years are now being crowded by scores of young trees (*Figure* 39). This has resulted in mature trees becoming stressed, as indicated by dieback at their tops (*Figure* 40). And not far from these giant sequoia groves, overcrowded red fir-dominated forests are also experiencing severe decline (*Figure* 41). Altogether, the dense forest cover, the many dead and dying trees, and the buildup of ladder fuels further results in increased likelihood of the entire forest succumbing to wildfire.[12]

Figure 38. Top dieback and death of mature trees from overcrowding by younger trees.

CHAPTER 7

Figure 39. An ancient giant sequoia surrounded by dozens of much younger trees, none more than two hundred years old. In other words, had I taken this photo two hundred years ago, only one tree would have been visible in this photo.

Figure 40. Top dieback due to overcrowding is also affecting ancient giant sequoias.

Figure 41. Dieback of overcrowded red fir-dominated forests in the central Sierra Nevada Range.

Cultural fire is important in controlling many of the pathogens and insect pests affecting plants. Native People have long known this, and they used fire to mitigate diseases and pests. In the words of Klamath River Jack (1916):

> Fire burn up old acorn that fall to the ground. Old acorn on ground have lots of worm; no burn old acorn, no burn old bark, old leaves, bugs and worms will come every year... Indian burn every year just same, so keep all ground clean, no bark, no dead leaf, no old wood on ground, no old wood on brush, so no bug can stay to eat leaf and no worm can stay to eat berry and acorn. Not much on the ground to make hot fire so never hurt big trees, where fire burn.[13]

By excluding the cleansing effects of fire, pathogens such as

CHAPTER 7

Sudden Oak Death and various insect infestations are now spreading quickly through the forests. Considering, too, that these outbreaks are occurring in the context of already overcrowded conditions, it is no wonder that so many diseases and pests are thriving in the absence of cultural fire regimes.

A comparable situation is occurring in the fire-suppressed forests of Australia, which grow in areas with Mediterranean climates similar to California's.[14] Early in 2010, I was contacted by Vic Jurskis, a scientist with Forests NSW (New South Wales, Australia), alerting me to the work he was doing on fire suppression and the decline of Australia's red gum and white cypress-pine forests. He shared with me several of his peer-reviewed papers that document the cultural burning of these forests by the Aboriginal People in the past, and the critical effects that fire suppression is now having on them. As he observes:

> Exclusion of fire and/or grazing has contributed to shrub or sapling encroachment, weed invasion, loss of herbal diversity and [forest] decline compounded by pests, parasites and diseases. The ancient trees that were established before European settlement are especially vulnerable because they have become weaker competitors for water and nutrients, whilst they are more vulnerable to fires because they typically have exposed dry wood that is easily ignited and burns readily.[15]

The above description of the importance of Aboriginal tending of the land and trees on the opposite side of the world is profoundly relevant, as these are also nearly the same words I would use to describe the mortality of oak forests in California!

Supported by substantial evidence that cultural burning can bring our native forests back to health, there has now emerged an entire discipline of fire science dedicated to this cause, populated largely by Indigenous fire experts, fire ecologists, fire fighters/lighters, and land managers.[16] But current recognition of the benefits of cultural fire by Western scientists has only arrived as the culmination of much hard work by some notable twentieth-century scholars.

Among these pioneers was Aldo Leopold, a noted American scientist and naturalist. In a letter to the ecologist Stanley Temple on November 19, 1938, he wrote, "As for the Indian-burning, we must concede that he certainly did know what he was doing, and that the product of his 'stewardship' was superior in many ways to that of the white man who displaced him." Leopold later stated, "But we have yet to become students of the land, to learn from the natives, and to recognize that the Indian in his way of burning was a more scientific land manager than we are in our way of plowing."[17]

In the 1950s, the well-known geographer Carl Sauer and his student Omer Stewart also attempted to raise awareness of how Native American cultural burning practices were critical for the creation and management of healthy forests and prairies.[18] In 1954, Stewart even wrote a book then titled *The Effects of Burning of Grasslands and Forests by Aborigines the World Over*. At the time, however, he chose not to publish it because of pushback from many ecologists of his day. It wasn't until 2002 that the book, retitled *Forgotten Fires: Native Americans and the Transient Wilderness*, finally appeared, presenting all of his original manuscripts on Native American burning.[19]

The introductions to that later volume by its editors Henry Lewis and M. Kat Anderson offer valuable insights as to why some ecologists have yet to respond to our current understanding of Indigenous land management practices. In her ecological analysis, Anderson, for example, tartly asserts:

> If ecologists and environmentalists were to endorse the premise that Indians shaped the ecology of certain plant communities with fire, they would have to rethink the tenets upon which their wilderness philosophies are based and would have to face up to

the removal of Native Americans from wilderness areas as in at least some instances a grave ecological *faux pas* that would ultimately undermine the unique habitat types and biological diversity that they sought to preserve.[20]

I, too, am well aware of how the wilderness ethos is still deeply embedded in the "environmental preservation" culture, and am regularly challenged by well-intentioned folks who truly believe that it is sacrilegious for humans to modify wild places in any way — aside from trail maintenance.

A recent book, *Smokescreen*, by Chad Hanson has galvanized the debate around the efficacy of using fuel reduction and prescribed fire to mitigate severe wildfires. Despite scientific meta-analyses (i.e., statistical analyses that combine the results of multiple independent studies) which conclude there is "overwhelming evidence" for the effectiveness of these interventions,[21] Hanson claims, instead, that these studies are based on "mistakes," "lies," or outright "deceit." Hanson even frames opposing arguments as "the new climate change denial of the 21st century."[22] This John Muir inspired work is essentially an attack on logging, which is mentioned more than *eight hundred* times. It is certainly fair to criticize the logging industry and their impact on severe wildfires, but by then saying fuel mitigation is just another for-profit industry like logging and should not be trusted, and by declaring that prescribed fires are not cultural fires and are thus "harmful," is in no way consistent to what most of us know about forest health and fire ecology in California.[23] Native American burning is barely mentioned in the book, and I know of no Indigenous People who have spoken out in support of Hanson's views. John Muir's writings certainly helped guide many twentieth-century colonizers, including me, to see and appreciate the sanctity of nature and take actions to preserve it. Today, however, in the twenty-first century, we have a lot more knowledge from a variety of Western and Indigenous sources informing us on how to make environmentally ethical decisions in the management of public lands.

A timely work that encapsulates the efforts to promote wise fire management is the book *The Pyrocene* by Stephen Pyne.[24] Pyne examines the use of intentional fires, namely cultural and prescribed fires, to manage our forests. Both activities can reduce the wildfire hazard and boost ecological health; however, cultural burning performs an assortment of other tasks, such as clearing for habitation, hunting, regeneration of basketry materials, improving the productivity of food plants, controlling insects and diseases, and more. While the seasonal and diurnal timing of cultural burning depends upon the desired outcome, prescribed fire is mainly used to mitigate fuels and so is often done in late summer or fall when fuel loads are higher and drier. Pile-burning is also a form of intentional fire and is frequently used in conjunction with both prescribed and cultural fires. Pyne's discussion of intentional fire, and its importance, is to prepare the reader for how humanity's future with fire may look. He ominously predicts, "There is no precedent for what we are about to experience, no means by which to triangulate from accumulated human wisdom into a future unlike anything we have known before."[25]

My many direct experiences with fire have taught me to be respectful and vigilant when in the company of flames. I have undergone Fire Fighter 2 and Emergency Medical Technician (EMT) trainings and have worked as a paid and volunteer firefighter in Colorado, California, and Alaska (*Figure* 42). I have stacked, lit, and tended hundreds of burn piles and have also spent years preparing forests for the eventual arrival of fire. I believe that, in the future, the management of most forests in California will involve the use of "good fire," turning fire fighters into fire lighters.

I've tried asking the following question in more long-winded ways, however someone I know (I wish I could remember who) phrased it most succinctly — "Why are we fighting fires under the worst of conditions when we should be lighting fires under the best of conditions?"

Figure 42. Photo taken at 3 AM while fighting a wildfire at Browns Lake, Kenai Peninsula, Alaska (June 1979).

All this is to emphasize the key roles that fire plays in the ecology of native forests in California and elsewhere. As I have learned firsthand, forming a covenant with fire is a critical life skill for improving the health of one's self, the community, and the surrounding forests.

The effects of liming on tree health and soil fertility

Beyond the reintroduction of fire, there are a number of other actions we, as forest stewards, can take to improve the health of trees and soils. For instance, given that the loss of nutrients, especially calcium, in declining forest ecosystems is closely linked to soil acidification, it should come as no surprise that the restoration of declining forests affected by acidification is achievable with the addition of calcium-rich mineral fertilizers.

In Chapter 5, I discussed the evidence for the positive effects of (calcium-rich) shell middens on the growth and health of adjacent and overlying mature trees here in Big Sur and elsewhere, as well as the notable high diversity of other plant species on shell midden sites. The Esselen knew what many ancient cultures also knew: that lime fertilization improves soil fertility and plant growth.

Western science has also weighed in on this point. Meta-analyses have shown that liming improves soil fertility, leading to increases in crop yield,[26] tree growth,[27] and forest productivity.[28] While there are many relevant studies that could be discussed here, let me focus on those works that examine the long-term effects of liming on soil fertility and tree health.

In an eight-year study on Masson pine forests in China, researchers found significant positive effects of lime treatments on various measures of soil fertility (e.g., pH, calcium content, and aluminum content) and pine growth (e.g., canopy density, twig and needle length and dry weight, and fine root length). As they conclude, "in the acidified and declining Masson pine stand of Chongqing, southwest China, application of limestone powder proved to be effective for ameliorating the soil acidity, improving the crown health condition, and stimulating the tree growth; these effects increased as the amount of applied limestone powder increased."[29]

A similar but larger-scale liming experiment was done at the Hubbard Brook Experimental Forest in New Hampshire in the late 1990s. In the decades prior, both conifer and broadleaf forests there were in decline as pH and calcium (Ca) levels in the soils and streams were falling. Then, in 1999, amendments of wollastonite, a natural calcium-rich silicate mineral, were made to an entire forested watershed using a heavy lift helicopter at a rate of about one-half ton per acre.[30] After more than a decade of follow-up studies, the results indicate that this treatment was effective in restoring soil calcium, mitigating acidification, improving tree health, and reversing forest decline.[31]

In one such study on the health of red spruce at Hubbard Brook, G. Hawley et al., concluded: "For all crown classes combined, winter injury was significantly greater for red spruce on the reference [untreated] watershed than for spruce on the Ca-addition watershed."[32] And in a study of a sugar maple forest in the same area, S. Juice et al. reported: "Foliar calcium concentration of canopy sugar maples [in the calcium treated watershed] increased markedly beginning the second year after treatment, and foliar manganese declined in years four and five. By 2005, the crown condition of sugar maple was much healthier in the treated watershed as compared with the untreated reference watershed."[33] Another study of sugar maple forests at Hubbard Brook by B. Huggett et al., found:

> Coinciding with foliar Ca differences, trees exhibited a significant difference in crown vigor and in percent branch dieback among treatments, with a trend towards improved canopy health as Ca levels increased. Annual basal area increment growth for the years following treatment initiation (1998-2004) was significantly greater in trees subjected to Ca addition compared with trees in control treatments.[34]

Additionally, in a long-term study of hardwood forests in Quebec, where both acid and lime treatments were applied to stands of sugar maple, R. Ouimet et al. reported, "over the two decades following treatment, basal area increments of sugar maple increased by 138% in limed plots compared with controls, while it decreased by 25% in acidified plots."[35]

With regard to studies specifically on oaks, an experiment on stands of sessile and English oaks in France found that, after 27 years, the application of lime enhanced fine-root biomass both in the topsoil and the deeper horizons, as well as improving soil fertility in the limed stands compared to the untreated control stands.[36] A separate greenhouse study on sessile oaks reported that liming enhanced soil fertility, alleviated aluminum toxicity, improved foliar nutrition, and stimulated the fine root and shoot growth of oak seedlings.[37]

I have long suspected that acidification and calcium depletion were playing important roles in the mortality of oaks in California. In 2005, I published data evaluating soil calcium and pH in both diseased and non-diseased California forest sites and found that the soils from the diseased sites were consistently lower in both calcium concentration and pH than nondiseased sites.[38]

Together this research not only emphasizes the positive effects of soil liming, it also indicates that soil acidification has a significant negative effect on forest ecosystems, a point I've tried to emphasize throughout this book. Furthermore, these findings show that liming can improve tree health and soil fertility at the watershed scale. These results are highly relevant to the science of fire mimicry.

Another related factor may be the matter of the deep cracks often seen to be forming in the bark of certain oaks. The bark of a tree is the first line of defense against diseases and pests. Thus, these cracks serve as open wounds that can allow pathogens and insects to more easily enter the tree. Given evidence that oak bark has a very high calcium content (in some species twice that of nitrogen[39]), could the bark deterioration of oaks be related, in part, to calcium deficiency in the soils? This is surely a topic worthy of further investigation.

Wood ash, volcanic ash, and biochar effects on tree health and soil fertility

The application of ash and other fire byproducts (i.e., biochar) to forests has been shown to have beneficial effects on forest habitats. Burning of forests, whether due to prescribed fire, cultural fire, or wildfire, produces an abundance of wood ash and biochar that have observable effects on soil fertility and tree health. Studies show that, following a fire, mineral nutrient levels (e.g., calcium, magnesium, and potassium) and the pH of the soils increase significantly as a consequence of the deposition of wood ash.[40] In some forests in California, the pH of surface soils has been seen to increase threefold (becoming one thousand times less acidic) after burning. One study found, too, that wood ash produced by fire in oak woodlands has a notably higher calcium content than the ash from nearby conifer forests.[41] This is probably why many of us who burn oak in our woodstoves have noticed the improved health of the nearby trees beneath which we spread the ashes.

These days, however, I mainly use volcanic ash as this is a richer source of essential trace elements. An owner of a company trying to sell me a volcanic ash mineral fertilizer once told me that he started his business after reports of bumper agricultural crops in eastern Washington following ash deposition from the Mt. Saint Helens eruption. As we know, volcanic regions have some of the most fertile and productive soils in the world. Several studies have shown that in only a year or two after volcanic ash deposition, nearby tree growth, as determined by changes in growth-ring widths, increases dramatically.[42]

Biochar, the charcoal remains of biomass burning, is likewise becoming a popular amendment for improving soil health. The TEK literature describes the wide use of biochar in Indigenous agriculture across Africa, Asia, and the Americas. Biochar is a major component of highly fertile terra preta (black earth) soils in western Africa,[43] the central Amazon,[44] and Australia,[45] which were created and maintained by the local Indigenous Peoples over millennia. And there is also a long history of the production and use of rice straw biochar as a soil amendment in Japan, Korea, and Vietnam. As described earlier, ancient shell middens constructed by Indigenous cultures worldwide also support highly productive soils that are typically rich in ash and biochar.

Western scientists have also examined the beneficial properties of biochar. Meta-analyses of biochar use in agriculture and forest restoration show, robustly, that biochar improves soil fertility, microbial activity, crop productivity, and tree growth.[46] Biochar is also observed to help remediate soil organic pollutants.[47] A key feature of biochar is its high surface area, which aids in the retention of nutrients and water in the soil. However, biochar alone cannot provide a comprehensive set of nutrients, and performs best when infused with essential minerals and organic compost.

An additional benefit of biochar is that the carbon it contains is highly stable, and so serves as a long-term sink of atmospheric carbon, sequestering it in the soil for thousands of years.

More on moss and lichen interactions with trees

The control of mosses and lichens is another way to assist forest health. Having addressed this topic in depth in the previous chapter, I would only add here that the buildup of mosses and lichens on trees adds significantly to the fire danger in a forest, not only for the ill health they promote, but also for their high flammability when dry.

Not surprisingly, one of the key reasons for the heavy buildup of mosses and lichens in California forests is the lack of fire. Mosses and lichens are known to be extremely sensitive to flames, ash, and pollutants from smoke. One can easily see that trees in areas that are regularly burned display little moss or lichen cover. Similarly, trees growing in urban areas with poor air quality rarely support an overabundance of lichens and mosses. The presence of mosses and lichens is, in fact, often seen as an indicator of "good" air quality. But this may be too good for some trees. Several Native elders, including Little Bear and Ron Goode, have told me that smoke in the canopies is healing for trees. Little Bear says that the smoke "sages" the trees, and he often builds fires beneath or near the canopies of large oaks.

Diseases, insect pests, and parasites

Besides preventing the buildup of mosses and lichens, fire also controls the level of pathogens (disease-causing organisms) and insect pests. Nature has devised an effective mechanism for keeping these aggravating organisms in check. As trees become sick and die from elevated levels of pathogens and pests, the fire danger becomes so severe that when fire does arrive, the diseases, pests, and many of their hosts are incinerated in the flames.

For instance, in a study published in 2005 examining the relationship between fire history and the incidence of the Sudden Oak Death (SOD) pathogen (*Phytophthera ramorum*), researchers found that SOD infections were "extremely rare" within the perimeter of any area burned since 1950.[48]

The Big Sur Basin Complex fire of 2008 gave SOD researchers a rare opportunity to further assess the effects of fire on previously monitored redwood/tanoak and mixed evergreen forests with and without symptoms of SOD. They found in the stands where, in 2006 and 2007, the trees had recently died from SOD, the "composite burn index" was significantly higher than in those plots without SOD.[49] Furthermore, they found significantly greater damage to the substrate strata of diseased vs. nondiseased stands, due to the high abundance of downed trees in infected stands. When combining all the strata of both recently SOD infested and past SOD infested sites, however, overall burn severity was not significantly different at the SOD sites compared to the non-SOD sites.[50] These findings suggest that diseased trees, at least those which have recently died, add considerably to the severity of wildfires (*Figure* 43).

In a related study, SOD researchers found that the Big Sur wildfires of 2008 suppressed, but did not eradicate, SOD from areas of the landscape that were previously heavily infected.[51] This observed short-term suppression makes me wonder how the SOD pathogen might respond to repeated cultural or prescribed fires over several years.

Perhaps not too surprising is the finding that SOD-infected trees can cause collateral damage to nearby redwoods. While redwoods are not directly compromised by SOD, the infected trees around them create extreme fire conditions that the redwoods sometimes cannot survive.[52]

Figure 43. Mature coast live oaks affected by Sudden Oak Death disease and oak borers in Big Sur.

Fire is also known to lower the incidence of root rot diseases that often affect oaks by promoting the growth of other fungi that inhibit these deadly pathogens.[53]

Insect pests, too, are affected by fire. A time-series study in western Montana found that the duration and intensity of spruce budworm outbreaks in Douglas fir and ponderosa pine forests has increased in areas with low fire frequency. This was attributed, in part, to the increase in host species (Douglas fir) brought on by fire suppression.[54] Studies of mountain pine beetle in burned, thinned, and untreated stands of ponderosa pines in Montana and Jeffrey pines in California, have shown that stands which have undergone both thinning and burning treatments exhibited significantly lower rates of pine beetle infestation than untreated control stands.[55]

There are, ironically, also circumstances where burning (with no thinning) has been shown to increase bark beetle activity on the surviving trees. While I have observed bark beetles infesting trees that survived a single catastrophic fire, it remains unclear to me how these surviving bark beetles would cope in the long term after being subjected to more frequent, cultural fires.

Another relevant example of using fire to control insect pests in oaks is provided by M. Kat Anderson, who has recounted the stories of members of the Yokut and North Fork Mono Tribes, which traditionally used cultural burning beneath the oaks to control acorn worms.[56] And in a later work she went into greater detail on the use of cultural fire to control diseases and insect pests. As she observes, "Indigenous people realized that the application of fire to patches and stands of plants covering small to large areas was the most effective tool for biological control of these organisms."[57] H. McCarthy had previously noted how Karuk women reported that "The trees are better if they are scorched by fire each year. This kills disease and pests."[58] Not unexpectedly, a recent scientific study corroborates this TEK that prescribed fire reduces insect infestations in acorns.[59]

In the absence of burning, many oak trees are now experiencing heavy infestations of mistletoe. Mistletoe (*Phoradendron* sp.) is a parasitic, evergreen plant that thrives on certain oaks and other trees, especially those that are stressed. Stressed oaks have less foliage to shade the mistletoe which, while parasitic, still needs light for photosynthesis. Why, then, am I not surprised to find that the smoke from fires is reported in the TEK literature to be an important control of mistletoe?[60]

Besides the proliferation of pests and pathogens affecting trees in fire-suppressed areas, there is also evidence that a rise in tick populations across the U.S. is tied to a lack of fire. The recent increase in both the number and incidence of tick-borne diseases represents a significant health threat to people and to wildlife. Yet, burning has been shown to have both direct effects (heat death) and indirect effects (habitat loss), which together lower tick populations. As a result, efforts are underway to reform approaches to tick management by thinning the forest and applying prescribed fire.[61]

To summarize, there are broad consequences of both fire and fire suppression for forest ecosystems. Cultural burning does appear to be an excellent means of maintaining the health of trees and the fertility of soils. Fire suppression, by contrast, often leads to forest decline and dangerous wildfires. However, under no circumstance is a catastrophic wildfire beneficial to a forest ecosystem. The next chapter describes ways to avoid the detrimental impacts of fire suppression.

8

RETURNING THE GIFT: THE PRACTICE OF FIRE MIMICRY

Given the many negative consequences of fire suppression, a clear course of future action will be to reestablish the widespread use of fire in managing our forests. Yet many forests are now so overgrown that lighting them afire would spell disaster. In these situations, where burning is impractical, the sensible approach is to mimic the effects of fire to the best of our abilities. As I hope to show, fire mimicry is a safe and effective strategy to "return the gift" and help improve the health of forests in both populated and rural settings, at least until such time as we can safely return fire to the land.

What is fire mimicry?

Fire mimicry, simply put, involves a suite of methods and materials designed to emulate the effects of fire, as well as other land-care practices of Native Peoples.[1] The use of fire mimicry in forest restoration has been described previously in the book *Mimicking Nature's Fire: Restoring Fire-Prone Forests in the West*.[2] In this work, Stephen Arno and Carl Fiedler detail efforts at fuel reduction through clearing of the woody understory and thinning of the young trees. They even discuss prescribed fires as a useful management tool, which, where possible, should be the eventual goal of fire mimicry practices. Similarly, a 2009 paper by J. Long describes using forest management activities such as clearing, thinning, and grazing to "emulate natural disturbance regimes."[3]

While both of these publications mention the agency of Native Americans within historical fire regimes, the authors' emphasis on mimicking "nature's" fire and emulating "natural" fire disturbance seems, to me, to represent an erasure of Native Peoples' role in the wise management of their lands. This is why I believe it would be better to emulate the effects of cultural burning rather than natural fires.

What I've learned and shared above about Native American tending practices, the succession of forest ecosystems, the problematic role of acidification in soil fertility, the beneficial effects of mineral amendments to forest health, the harms of excess moss and lichen cover on trees, and the desire to apply this information to help address the issue of oak mortality is condensed here into a set of forest restoration practices that mimic healthy cultural fires. These practices fall into three broad categories: 1) habitat care: clearing, thinning, and pruning to reduce competition and ladder fuels; 2) root care: treating the soils with calcium-rich mineral fertilizers, compost tea, and biochar; and 3) trunk care: controlling mosses, lichens, diseases, and insect pests to the extent possible. The following sections present the detailed protocol I've developed for tending trees in this way.

Establishing a connection with each site and each tree

As with any practice there is an art and flow to the preparation process, which in this case starts with finding a personal connection between the trees and those responsible for their care. For me, this connection develops through initial introductions and an inspection of the place. I quiz the property owners about how long they've lived at the site and any observations of changes they've noticed in their trees. Oftentimes they show me stumps of trees that have died and been removed. Afterwards, I instinctively wander towards each tree, identify it, and assess its health by observing the density of its canopy as well as the degree to which it is overgrown by understory brush and young trees. I inspect the trunk of the tree for stem canker infections and insect infestations, and then sometimes, sadly, I must inform some owners that a tree cannot be saved. All this helps me to get my mind around the scope and planning of the project.

At the start of the workday, I do a "walk through" with my associates. We visit all the trees to tend, highlight those trees in need of extra attention, and determine safe and efficient ways to navigate the site. This is our time to feel, appreciate, and say a quiet prayer for the trees. My humble mantra is: "When I care for the trees, the trees care for me!"

Habitat care: Clearing, thinning, pruning, staging, and uses of cut materials

Invariably, the first steps of fire mimicry in an overgrown forest involve preparations for cutting much of the woody understory, thinning the young trees, pruning the remaining mature trees, and staging the cut materials for processing (e.g., burning, mulching, and/or cutting for firewood). These steps require a set of tools that include several small chainsaws, a large chainsaw, a power pole saw, a hand pole saw, loppers, machetes, and rakes. Safety is paramount, so we use gloves, chaps, safety glasses, ear protection, head gear, and skin cover to protect from poison oak. The supplies we have on hand include two-cycle fuel, bar oil, a chain wrench, sharpening files, extra chains, tarps, and flagging. I always stage my tools and supplies on ground tarps to keep everything as clean as possible.[4]

Next, I locate and flag the native trees or shrubs that we want to preserve. For instance, here on the Central Coast of California I try to minimize removal of toyons, hazelnuts, silk tassels, black sages, white sages, elderberries, gooseberries, and other culturally important native species. Preserving snags is recommended in certain circumstances. Large standing dead trees can provide excellent wildlife habitat after removal of the outer branches, but only if they are in places where, upon failure, the tree would not damage any nearby trees or infrastructure. On steep slopes I might want to leave certain young trees that may be beneficial in keeping the slope stable. I also note any bird nests and avoid working within three hundred feet of these sites (or within five hundred feet if they are raptor nests) during the spring nesting season.

If it is fire season, which can be any time of year in California, it is important to have a nearby source of water in case of accidental fire, as sparks from a chainsaw can sometimes ignite adjacent vegetation. If possible, I'll attach a hose to a water source that can reach the clearing site; otherwise, I bring portable backpack sprayers filled with water.

At this point, I take photos of the site to document its appearance before the work begins. Sometimes, I even set up a time-lapse photography station to arrive at a better visualization of the clearing process.

All the above steps are taken in preparation for the moment when our bodies and tools become surrogates for fire. To begin, we cut most of the woody vegetation at ground level, remove many of the young trees, especially any invading pines, firs, and bay laurels, and prune the lower branches of the mature trees. As the before-and-after photographs show, this typically transforms densely overgrown hillsides from a condition where the trees can barely be seen to one where all of them are suddenly visible (*Figure* 44).

CHAPTER 8

Figure 44. Before and after photos of clearing, thinning, and pruning around oaks and redwoods in Big Sur. Photo dates are February 1, 2014 (left) and February 20, 2014 (right). Note the abundant oak firewood being recycled in lower right image. See *Figure* 54 for later photos of these same locations.

Clearing, thinning, and pruning work needs to be done roughly in that order. Brush must first be cleared and removed to reach the young trees; then the young trees need to be cut and removed to make room to effectively prune the mature trees. My YouTube channel has dozens of time-lapse and real-time videos showing these and other tasks involving fire mimicry treatments.[5]

A word of caution for anyone wanting to do this work in Big Sur and beyond, poison oak grows widely here on the Central Coast and is extremely dense in places. I and most of the crew have almost no reaction to poison oak — it is our superpower! (Some of the crew even do side gigs removing poison oak.) But if you are sensitive to poison oak, or don't know, this kind of work is probably not for you. I've seen people have serious reactions to poison oak that sometimes require medical care. Too, if you do choose to work in it, be aware that you can readily transmit poison oak to others until you remove your clothes and wash up.

There are some slightly different guidelines with regard to implementing these tasks around mature trees in landscaped and garden settings. In these instances, pruning and removal of landscaped shrubs and young trees growing close to the mature trees is sometimes necessary. But in many situations this is not practical. In those cases, I try to ensure that the mature trees are cleared of any vines and any leafy vegetation within two to three feet of the trunk.

Ideally, most or all of the cut material should remain on site, and the movement and proper staging of the cut material is, at times, as much an effort as the cutting itself. Decisions need to be made beforehand as to whether the material will be mulched, burned for ash and biochar, processed for firewood, and/or kept for terrace construction.

If it is to be mulched, then the material needs to be staged near access to a road. Once mulched, this material can be used to cover any barren areas around trees, which will reduce moisture loss from the soil. If it is to be burned, then the material should be piled adjacent to safe burn sites and left to dry. Larger rounds intended for firewood or in terrace construction should be piled separately.

While in certain instances materials can be piled and later burned in place, more often the safer procedure is to prepare a cleared site to start a fire, then steadily feed it from the nearby slash piles (*Figure* 45). Burning should never be done without a water source, which is essential not only for safety, but also to produce biochar.

The simple process of putting out a pile burn with plenty of water and shovels can be used to create abundant ash and biochar that can later be spread around the nearby mature trees. We've even used biochar kilns to increase the quantity of biochar production. The remaining ash contains alkaline-rich mineral nutrients, and the biochar acts to bind water and nutrients in the soil, which can be then incorporated into the next phase of fire mimicry.

CHAPTER 8

Figure 45. Pile-burning of cleared materials at Indian Canyon.

Root care: Mineral nutrients, biochar, and compost tea

When looking at a majestic oak, with its dense spreading canopy reaching to the sky, one should always be reminded of the nature of reality — as above, so below. For every branch visible above ground there is a corresponding network of invisible roots below ground. So too, when I see dieback in the upper and outer branches of a tree, I sense a problem with the roots. Roots are where the trees obtain most of the water and nutrients needed to sustain the aboveground stems, branches, and foliage. If water and nutrients are not available in the soil, the roots will wither and die, as will the leaves and branches. Thus, the focus of this next set of tasks is to improve soil fertility and root health.

The first step in the nourishment process is to amend the soils with alkaline-rich mineral nutrients. As mentioned earlier, without fire, soils become acidified, which accelerates the leaching of mineral nutrients including calcium, sodium, magnesium, potassium, and a host of trace elements. Fortunately, there are a range of natural products that can be used to prepare an effective mineral nutrient fertilizer.

The formulation I prefer is an approximately fifty-fifty mixture (by weight) of oyster-shell flour and a micronized trace mineral fertilizer (Safety tip: dust masks should be used whenever working with fine powdery materials). The oyster-shell flour provides the essential alkaline nutrients (e.g., calcium, magnesium, sodium), and the trace mineral product provides dozens of important micronutrients to the soils. While there are several powdered rock products rich in trace minerals that can improve soil fertility and plant health, none seem to be as effective as micronized AZOMITE®, which comes from an ancient volcanic ash deposit in central Utah.[6] Lore has it that the local Utes traveled to the pink rock formation to quarry the stone, crush it, and apply it to the soils before planting maize.[7] Following their illegal land grabs, Western colonizers also started using the crushed pink rock on their crops and they, too, observed its beneficial effects.

None of the trace mineral products alone have sufficient calcium content needed to properly amend the soils around sick trees. That is why I add shell flour to the mix. Oyster-shell flour is an excellent natural source of calcium that is commonly used both as a soil fertilizer and as a feed additive in the livestock industry. Oyster-shell flour and AZOMITE® also have no toxicity and are both are approved for organic agriculture.

When fertilizing soils, I also add biochar to the above mineral mixture. I use about one to two cubic feet of finely crushed biochar/wood ash per one hundred pounds of minerals. For some reason, commercially available biochar is quite expensive, so I've been relying on biochar from burn piles, which produce some biochar, but this is often not enough to fertilize whole stands of sick trees.

Once the mineral/biochar mixture is prepared, I put it in a bucket and use a hand scoop (ice or grain scoops work well) to apply it evenly to the soil surface beneath the entire canopy of a tree, as far as or even beyond the dripline (*Figure* 46). For a typical mature oak, I will use about thirty to fifty pounds of biochar/minerals, and for a very large oak I might use eighty to one hundred pounds. Application can result in some fine dust particles in the air, so wearing a dust mask is recommended. Afterwards, I water the minerals and biochar fragments into the soil. This fertilization treatment can be done any time of year, as the mineral nutrients and biochar are relatively stable and will not degrade in the soil even during prolonged dry periods.

Figure 46. Application of alkaline minerals and biochar mixture around a mature oak at Indian Canyon.

Next comes the application of compost tea. Compost tea is essentially an infusion of compost steeped in water for several days with the aim of transferring organic matter, beneficial microbes, and nutrients into solution. It has been shown to improve both soil fertility and plant productivity across a range of environments.[8] The compost tea I brew includes biodynamic compost, AZOMITE® as an additional mineral nutrient source, and a bit of molasses to help stimulate microbe production. Following application, the compost tea binds with the soil particles and biochar, inoculating them with spores of beneficial bacteria and fungi that, along with other soil fauna and flora, convert the mineral nutrients into forms that are available to the trees. While the application of compost tea is not really mimicking fire, it is an easy additional task to ensure good tree and soil health.

I prepare the compost tea in five-gallon buckets, using one pound of biodynamic compost, two tablespoons of molasses, one tablespoon micronized AZOMITE®, and the remainder water. The water cannot be chlorinated, or it will kill the microbes we are trying to culture. If the only available water source is chlorinated, I either fill a bucket with the water and let it stand in the sun for a day or bubble it with an aerator. Once the chlorine has been released, the water can then be used to brew the tea. This involves mixing the ingredients and inserting a fish-tank aerator to keep the brew oxygenated for several days. Upon completion, the buckets may be sealed with watertight lids and transported to the field.

In the field, I strain the compost tea through a fine mesh sieve into garden sprayers and apply it to the entire soil surface beneath the canopy of the tree, from the trunk to the dripline. Most any garden sprayer will do, although I prefer the two-gallon hand-held type, because they are lightweight, simple to use, and east to maintain. Typically, I apply about one to two gallons of compost tea per mature tree. The compost tea can also be applied to the leaves and branches of sick fruit trees, often with surprisingly positive results.

Notice that the bulk of the soil amendments have little nitrogen (N) or phosphorus (P) content. I have found that adding these nutrients can acidify the soils if not applied properly. The compost tea provides some of N and P, but in my view most of the sick trees I encounter are, arguably, more deficient in mineral nutrients than in nitrogen or phosphorus.

After the application of the mineral, biochar, and compost tea fertilizers, the soils are again generously watered to speed the activation of the nutrients. There seems, however, to be some disagreement among tree-care experts in California on the watering of oaks during the dry season. Some say it should not be done, as the native oaks are adapted to our long, dry summers. Others, including myself, do not see any harm in watering oaks once or twice a month during the dry season, particularly when in a drought. However, continuous watering all year long, as sometimes happens in lawn and garden areas, is discouraged as it often leads to moss buildup, leaching of soil nutrients, and/or root rot disease. In my practice, those clients who are able to give their oaks a few waterings during the dry season see the best results from the fire mimicry treatments.

Trunk care: Controlling mosses, lichens, diseases, and insect pests

Perhaps the most difficult task of fire mimicry is dealing with trees that have a heavy growth of mosses and lichens in their canopies, and/or are being attacked by diseases and insect pests. While fire mimicry can address issues at the ground and soil level, it cannot replicate the smoke and flames that would normally control these harmful organisms higher in the canopy. Thinning and clearing of bay laurels around oaks helps significantly to reduce the spread of Sudden Oak Death disease, because bay laurels are the primary vector for this pathogen. But for the most part, the problems posed by the overabundance of mosses, lichens, diseases, and pests will not be solved at the landscape scale until cultural fire is reintroduced into these forests.

For now, the focus of moss, lichen, disease, and insect pest control in fire mimicry must be near the base. The trunk of the tree is of utmost importance in its health. The reciprocal exchange of resources between the leaves and the roots must all pass through the trunk, so anything that can be done to improve the health and integrity of the trunk is crucial.

Trunk care starts with the removal of mosses and lichens using a wire brush (and wearing a dust mask), from the ground to a height that can be easily reached, typically five to six feet (*Figure* 47). Not all the mosses and lichens need to be eradicated, just the bulk of them; any that remain will be mitigated by the limewash procedure that follows.

Figure 47. Removing heavy moss cover from the trunk of a valley oak using a wire brush.

Once the mosses and lichens are removed I carefully inspect the newly exposed trunk for stem canker infections or insect infestations that may have been hidden by the heavy moss and lichen cover. If any infections or infestations are found, and are deemed operable, then a surgical procedure is performed — the details of which I will describe later in this chapter. If no infections or infestations are found, a limewash (aka whitewash) is then prepared and applied to the trunks of the trees. Before I go into the details of this procedure, however, let me explain more about the history and traditional use of limewash on trees.

The historical use and effectiveness of limewash

The limewashing of trees is a custom that probably goes back thousands of years, and it is still practiced in various cultures around the world. We know that ancient Egyptian, Middle Eastern, Mesoamerican, Indian, and Chinese cultures all used limewash and/or lime plaster on their buildings, and limewash was widely used on both brick and wooden buildings in the U.K. and U.S. in the seventeenth, eighteenth, and nineteenth centuries.[9] While I have little doubt that trees have been treated since the invention of limewash, the earliest record of limewashed trees I'm aware of are historical photographs of painted trees in China dating from the mid-1800s.[10] The 1888 painting *The White Orchard*, by Vincent Van Gogh, likewise depicts trees covered in limewash.

My father, Frank Klinger, once showed me pictures of oaks they use to limewash on his childhood farm in southern Ohio in the early 1900s, saying that he thought they did it to make the trees "look pretty." And who among my fellow travelers to China, India, Nepal, Indonesia, the Congo, Brazil, Mexico, Costa Rica, Poland, Russia, etc., have not wondered why the trunks of so many trees in these and other places are painted white? I've recorded limewashed trees surrounding the Taj Mahal in India, lining the entrances to five-star resorts in Mexico, around government buildings in Russia, and in the famed Stone Forest of China (*Figure* 48).

Figure 48. Limewashed tree in the Stone Forest, a geological wonder in Yunnan Province, China.

In the past, limewash was commonly used on structures and trees to improve their durability and longevity. If you ask the locals why they still paint limewash on trees, as I have on numerous occasions, they typically say the limewash reduces insect pests and diseases, as well as providing protection from the sun. In Russia, they claim that the limewash also discourages bark damage by deer and rabbits trying to feed on the tree's cambium. And many locals point out that the limewashed trees along roads tend to be more easily seen and avoided by motorists, extending the health and longevity of both the trees and the people. Some have even told me that insect pests are readily seen against the white background of the limewash and are thus more thoroughly predated by birds. The high salt content of limewash also discourages snail pests that will feed on young leaves and fruits of certain trees. It is apparent that a range of cultures worldwide, especially those of the Indigenous

CHAPTER 8

Peoples, are aware of the benefits of limewash and continue to use it widely.

Another effect of limewash, which was shown to me years ago by forest managers in a remote part of China, is the mitigation of mosses and lichens growing on the trunks. They explained that the limewashed trunks do not allow mosses and lichens to grow, while non-limewashed sections of the same trunks have a heavy moss and lichen cover (*Figure* 49).

Figure 49. Limewashed tree along a roadside in Yunnan Province, China. Note the dense growth of mosses on the upper portions of the trunk without limewash.

Limewash is still commonly employed as a highly breathable, antiseptic, antifungal, odorless, and nonallergic paint derived from a solution of hydrated lime, salt, and various binding agents. It is known to be a fire retardant and when applied to wood, can slow deterioration due to rot and weathering. Limewash is still widely used on barns and other structures of the modern dairy and poultry industries to promote healthier, anti-bacterial conditions for livestock. According to the National Park Service, "Limewash can eliminate mosquito larvae, reduce odors where animals are kept, and when painted on roofs it reduces inside temperatures up to 10 degrees… Painted on tree trunks, limewash prevents disease, sunburn or frost injury." These words describing the use of limewash appeared as part of an exhibition at an old plantation at Cane River Creole National Historical Park in Louisiana. They add that "limewashing the bases of trees' trunks occurred regularly into and throughout the early 20th century."[11] Interestingly, the exhibition also reported that the hydrated lime used in preparing the limewash was sourced from local shellmound deposits!

While anecdotal evidence abounds for the efficacy of limewash in improving tree health, there are few controlled studies of limewash treatments to be found in the scientific literature.[12] The earliest study I have found is from 1898. In it, various trunk alkalizing treatments were tested on rows of peach trees for control of peach borers. These included the application of hydrated lime plus milk, hydrated lime plus water, and even a mixture of Bordeaux (wine) and hydrated lime. All of these experimental treatments proved effective, reducing the number of borers from an average of eight per row prior to treatment to zero afterwards, compared to the control row of peach trees where an average of two borers were found both before and after treatments. I find it interesting that the study reported the hydrated lime plus milk mixture exhibited the most durable cover (see limewash recipe below).[13]

In a more recent study on the efficacy of limewash on sick common fig trees in California, the researchers conclude that, "Treatments with whitewash [traditional limewash] seem to be effective at both protecting trees from sunburn, thus preventing development of cracks and other wounds on shoots, and suppressing infection by the pathogen and canker formation."[14] Considering the prevalence of stem canker infections in many sick California oaks, this study showing that limewash can suppress canker infections in figs is highly encouraging. Another controlled study comparing limewash treatments to untreated trees in Golden Delicious apple orchards in Mexico showed that the limewashed trees had 25 percent greater budbreak and 27 percent higher fruit yields than the untreated control trees.[15]

Despite the dearth of experimental studies, many state agricultural services in the U.S. recommend limewash application on various crop trees to reduce sunscald and insect activity. For instance, the Ohio Agricultural Experiment Station recommends three limewash applications (April, July, and September) to control for bark beetles in peach and cherry trees.[16] The Oklahoma the Cooperative Extension Service recommends limewashing pecan trees to control weevils.[17] In Oregon, the State Agricultural Extension Service recommends applying limewash to orchard trees to mitigate against sunscald and winter injury.[18] And limewash treatments to prevent sunscald of trees are also recommended by the agricultural extension services in Tennessee, California, New Mexico, and Arizona. State entomologists investigating leafhopper damage in California orange groves further recommend limewash treatments to the entire tree.[19]

Limewash preparation and application

There are any number of limewash or whitewash recipes that are said to be effective in maintaining healthy trees. Searching the internet for "limewash" or "whitewash" recipes will bring up scores of results, each with similar formulations. The common ingredients in most limewash recipes are powdered calcium hydroxide $(Ca(OH)_2$; also known as hydrated lime, type S lime, or miracle lime), salt, milk, and water. The lime component imparts the antiseptic properties, the milk and salt assist with adhesion and durability, and the water provides a soluble medium. Other traditional recipes have included volcanic ash, wood ash, casein, hide glue, linseed oil, rice flour, soap, molasses, and even goat's blood. Some recipes call for latex paint, but I would avoid this and any other ingredients that are not suitable for organic agriculture. Besides, I'm fairly certain that the mineral nutrients provided by limewash are far more beneficial to the health and metabolism of the trees than latex paint.

I have tried a few different limewash recipes over the years and have settled on one that involves inexpensive, readily available ingredients and is easy to prepare. My current recipe is as follows:

- 1 gallon of organic whole milk
- 1 cup fine sea salt
- 2 cups micronized AZOMITE®
- 18 *heaping* cups of hydrated lime
- 3 to 4 gallons of water (depending on brush or spray application)

Wearing a dust mask, the above ingredients are combined in a five-gallon bucket. If the limewash is to be applied with a brush, then I prefer the limewash to be a bit thicker, and only add enough water to fill the bucket to the four-gallon level. More often, however, I wish to spray the limewash, so I add water to the five-gallon level. I then use a variable-speed handheld power mixer to blend the solution for about three to five minutes. For spray application, I strain the mixture through a fine mesh sieve (mesh size = 1/32 inch) and pour it into a standard handheld garden sprayer, *modified by removing the filter.* Once prepared, I spray or paint the limewash onto the entire surface of the trunk, from the root crown to a height of five or six feet for mature trees, and three to four feet for small trees. It is important to work the limewash into the cracks of the bark, as these are areas where minerals will be most protective, and where they can be absorbed into the tree more quickly.

If limewash is being applied in gardened landscapes, then I use lightweight tarps to cover any nearby plants, fences, or walkways around the trees. While the limewash is not harmful to the plants, the overspray can cause unsightly white spots on the leaves and surrounding landscape elements.

I have sometimes run into concerns with certain clients who do not like to see their trees painted white as it makes them appear "unnatural." I understand their concern, but I remind them that "natural" here in California would mean burning around their oaks every few years. The limewash, which I sometimes call "liquid fire," has many of the properties of actual fire, and it needs not be viewed as unaesthetic. For me, it's a sign to my neighbors that I am taking care of my trees.

Still, I must admit to occasionally caving to their requests, especially if it might otherwise involve no treatment at all for an ancestor oak. In such cases, I have made a nontoxic tinting agent from a natural black iron-oxide pigment.[20] Mixed with water, it works well with oaks and certain other trees. However, I only apply the tint *after* the limewash has dried on the trunk. I have tried mixing the tint into the limewash but cannot achieve a natural bark color or texture this way. Oftentimes, the tint needs to be reapplied a few weeks later to get acceptable results for fussy clients.

One last point about the importance of the white and porous nature of limewash. If any treated trees have canker infections that go undetected upon initial inspection, the dark ooze from these infections will eventually flow to the trunk surface and be readily seen against the white limewashed background.

Surgical procedures

I often draw upon the analogy that some of the fire mimicry procedures are like good dental care — brushing the mosses and lichens off the tree reduces the chance of decay, and applying an antiseptic limewash, much like a toothpaste, inhibits decay-causing organisms. But sometimes our teeth do suffer decay. In that event, a dentist will remove the decayed tissue, sterilize the cavity surface, and apply an antiseptic filling.[21]

The surgical protocols I use on canker disease infections in trees are guided by the same goals as tooth cavity repair: removal of infected tissue, sterilization of wound surface, and application of an antiseptic covering. While the use of surgery to remove stem canker infections in stone fruit trees is known to be highly effective,[22] there is, surprisingly, nothing in the literature about stem canker surgeries applied to oaks or other trees. Apparently surgical attempts at canker removal in oaks have not occurred, or, if so, they have not been reported.

The surgical procedures below have been developed and implemented mainly on coast live oaks and differ slightly from those used on fruit trees. The efficacy of this surgical practice beyond oaks is not known, but I would welcome intrepid tree tenders to try these methods on other kinds of infected trees and document the results. Otherwise, I hesitate to encourage folks to engage in taking cutting tools to infected trees without proper guidance. At the same time, I realize that finding that guidance will not be possible for many readers, so I'm sharing the details in hope that some will find them useful.

When an oak becomes infected with a stem canker disease, be it Sudden Oak Death, *Armillaria* sp., or other pathogen, there is usually some visible bleeding associated with the infection (*Figure 50*). The bleeding comes in the form of a thick black or reddish-brown fluid oozing out of cracks in the bark. This ooze is a typical defense response of trees trying to push out diseases and insect pests. The cracks themselves can also be an indicator of an infection. Most deep cracks in the bark of an oak run vertically along the trunk, but if they run horizontally, this is often an indication the tissue underlying these cracks is dead. Insect infestations can also cause bleeding and may be identified by numerous bore holes in the bark, oftentimes surrounded by frass exuded by certain pests such as termites and bark beetles.

CHAPTER 8

Figure 50. Bleeding stem canker infection in the trunk of a coast live oak.

Unfortunately, in many cases a tree may be too severely infected to warrant surgery. These are situations where the canker has already girdled one-third or more of the trunk. The tree may also have a systemic infection, where numerous cankers occur at various places on the trunk. This, too, is not easily remedied by surgery.

If an infection is found that is deemed operable, I prepare my equipment and materials. These include a large hand axe, a small hand axe, a power angle grinder with cutting blades, a power multi-tool, a propane welder's torch, a fire extinguisher, a hand broom, a ground tarp, a paint brush, mineral poultice, gloves, and safety glasses.[23] All the bladed tools are cleaned with alcohol and the cutting surfaces sterilized with a blowtorch beforehand. The surgical site is then prepared by laying the tarp on the ground beneath the infected area to gather the diseased tissue as it is removed.

Focusing on the places of the trunk where the tissue is clearly bleeding, I start swinging my axe (*Figure* 51). I try to make each cut calculated and precise, often taking a moment between swings to inspect my progress. My first goal is to find the boundary between the infected and healthy tissue. A rule of thumb about canker surgeries is that the infections are nearly always larger than what can be observed before cutting into the trunk. I've learned not to be timid when I see the infections extending beyond my initial cuts, and continue cutting back until encountering the "triple edge" of the canker boundary — the interface of healthy reddish bark tissue, dark-colored diseased tissue, and underlying whitish wood tissue.

CHAPTER 8

Figure 51. Surgical removal of a large stem canker starting with an axe.

After the infection is largely removed, I use my hand broom to clean any loose fragments and sawdust from the wound. I then gather the debris from the tarp and place it in a green waste bin or set it aside for composting.

Next is the sterilization/cauterization procedure (*Figure* 52). With a fire extinguisher or hose nearby, I use a hand-held propane torch to carefully burn the entire wound, making sure that the edges of the wound and everything else is scorched mostly black. Cauterization has several important functions: the heat can help kill any lingering bits of diseased tissue; it creates a charred surface that inhibits wood rot and other harmful organisms; and it stimulates a rapid healing of the bark tissue over the wound.

CHAPTER 8

Figure 52. Cauterization of the surgical wound using a propane torch.

Once I've delineated this boundary, I continue removing the larger pieces of dead and infected tissue within the remainder of the canker using both my large and small axes. After the bulk of the canker tissue is removed, I bring in my power cutting tools — a bladed angle grinder and a multitool. The bladed angle grinder is a bit of a beast and needs to be operated carefully, but with proper use it works well to remove most of the remaining diseased tissue. The multitool has a safe, reticulating cutting edge that allows me to cut out any lingering pockets of infected tissue.

A word of warning when cauterizing wounds — in certain instances I've encountered wood tissue which is rotted and punky. I see this most often with *Armillaria* infections. Soft, punky wood will readily catch fire and can be difficult to extinguish. This happened to me on one occasion, and it took me some time to stop the smoldering. I've since learned to recognize any rotted wood tissue and avoid burning it.

While the potential of smoldering tissue is a concern, the subsequent application of a wet, mineral poultice provides another preventive measure against accidental ignitions (*Figure* 53). The poultice is essentially what remains of the limewash that is too thick to go through the sprayer. These dregs are allowed to settle for several days before the excess is water poured off, leaving a thick mineral-rich poultice that may be applied with a paint brush to the entire surgical wound.

CHAPTER 8

Figure 53. Application of limewash poultice to the surgical wound.

After a few days' time, when the poultice has dried, there may be some weeping of dark tannins from the wounds. While this weeping can appear to be residual infections, most of the time it is just the tree's normal response to wounding. If there are any residual infections, and sometimes there are, these will exhibit continued weeping after about a year. At this point, a follow-up surgery to remove any remaining infected tissue is in order. As mentioned previously, videos of these surgical procedures and other steps of fire mimicry can be found on my YouTube channel.[24]

To conclude, stem canker surgeries are not for the faint of heart. I remember my first few surgeries, which were largely unsuccessful, using only an axe and chisel to remove the infections, fearing that I might be compromising the tree's health. I've now learned that if a surgery is attempted, it needs to be aggressive and remove all discolored, infected tissue. If an infection starts tracking away from the central infection site, it needs to be followed. Of course, there is a point where the wound becomes so large that it may be compromising the life of the tree. In such cases I will end the surgery and give the tree a year or so to recover, before attempting any additional surgery. Overall, these surgical procedures have clearly been worthwhile for improving oak health, as I will show in the results chapter that follows.

9
ASSESSING THE EFFICACY OF FIRE MIMICRY

Over the course of my work on trees, especially oaks, in California, I have made a special attempt to document the results, both to show their efficacy and improve my methods. This has largely involved a series of before-and-after photographs, in some cases following courses of treatment lasting more than ten years.

Repeat photography

Before delving into these photographic results, I want to describe the procedures I use to prepare them. I mainly follow the methods of repeat photography, where photos from one location are repeatedly taken over a period of time, preferably on or near the same day of the month, the same time of day, and under similar light conditions to document vegetation change. I learned this technique from my Ph.D. advisor, Dr. Tom Veblen, who has used it with historical photos to show rapid landscape-scale changes in forest cover due to fire suppression in the Colorado Front Range over the past century.[1] Besides trying to match the location, date, and time of day, successive photos should also attempt to replicate the exposure settings, the framing, and the lens angle as best possible.

My method begins with an initial set of photos of the individual trees to be treated on or near the date of their initial treatment. I prefer to photograph the trees around midday when the sun is highest and illuminates more of the canopy. For each photo I try to find a position where the sun is mostly at my back and the tree's canopy is displayed against a background that contains as much sky as possible, as a leafy background will obscure results. Moreover, I've happily learned from a few mistakes that it is important to frame the initial photos to allow for future upward and outward growth of the canopy, because in some cases the canopies of treated trees have outgrown the frames of my original photos.

After experimenting with various types of cameras, the equipment I use is a high-resolution digital camera and a zoom lens with a range of ~12 to 30+ mm. Geolocation is a helpful added feature. Cameras on mobile phones can be used effectively at times; however, in many cases I find there is a need for a lens angle significantly wider than that available on most of these devices.

Afterwards, I download these images to a specific case-study folder on my computer and assign each tree image a chronological case number. I then sync the folder to a tablet so that the images can be displayed readily in the field when seeking to reproduce their original location and angle of view.

With each treatment I try to take follow-up photos yearly at or near the calendar date and time of the initial photos. Holding my tablet in one hand and my camera in the other, I am usually able to precisely locate and recapture the scene of the initial image. These visits also allow for a ground-level inspection of the treated trees and surrounding vegetation. I then display the new photos side-by-side with the initial photos and cropped, if necessary. I usually also enhance the contrast of all photos by 20 percent to accentuate differences in canopy density.

CHAPTER 9

. . .

Before-and-after results of fire mimicry treatments

The most poignant part of the fire mimicry work, for me, is seeing the results of these efforts. There is nothing more rewarding than the sight of a sickly oak, which may have struggled for decades, coming back to health in just a few years.

The time-series results shown below highlight the effects of four aspects of fire mimicry: 1) changes in the forest subcanopy; 2) changes in tree canopy density and greenness; 3) recovery of trees that burned *after* being treated with fire mimicry; and 4) wound recovery following surgeries. While this evidence is based entirely on subjective assessments of the before-and-after photos, there is a lot of visual information in the photos that cannot be readily obtained by objective analyses. I leave it up to readers to make their own judgements about what they see. A full viewing of these and other fire mimicry results can be found in the archives at www.suddenoaklife.org.

Before-and-after analysis of our clearing, thinning, pruning, and pile-burning treatments on the forest subcanopy generally show that in subsequent years the growth of the woody understory growth is greatly reduced, while herbaceous groundcover increases (*Figure* 54). These images reveal a significant reduction in the fire hazard via the removal of ladder fuels, an opening of the forest canopy, and an increase in forage for wildlife within treated areas.

Figure 54. Before-and-after photos of clearing, thinning, and pruning around oaks and redwoods in Big Sur. Photo dates are February 1, 2014 (left) and February 10, 2018 (right). See *Figure* 44 for earlier images of these same locations.

The bulk of my results thus represent yearly assessments using repeat photography of the canopies of individual trees that have been treated with fire mimicry. In a few cases I have also obtained before and after photos of nearby "control" trees that have not been treated to help better interpret the results. Most of these results relate to coast live oaks, but I have also sought to document fire mimicry treatment of other species of oaks, as well as buckeyes, pines, and redwoods.

Somewhat surprisingly, I have found that many treated trees exhibit noticeable improvement in canopy health *after just one year*. *Figure* 55 shows case studies of oaks that have made remarkable one-year recoveries, even in the middle of a drought.

Figure 55. Two coast live oaks in Soquel, CA, that responded to fire mimicry treatments after just one (drought) year. The trees on the left were treated and photographed on September 27, 2020. On the right are photographs of these same trees on September 28, 2021.

More typically, trees take longer than one year to respond to treatments. *Figure* 56 are two ancestor coast live oaks on the mend after three years. And *Figure* 57 depict coast live oaks after five years and ten years of fire mimicry treatments. The observation that fire mimicry treatments can continuously improve the health of trees for up to ten or more years is a clear indication that these treatments are helping the oaks.

Figure 56. Two ancestor coast live oaks in Big Sur, CA, showing noticeable improvement after three years of fire mimicry treatments. The upper oak was diseased and was also treated with surgery. Images on left taken October 25, 2020. The images on right were taken October 23, 2023.

CHAPTER 9

Figure 57. Responses of Big Sur coast live oaks to fire mimicry treatments after five years and ten years. The upper photos were taken February 21, 2018 (left) and March 20, 2023 (right). The middle photos taken March 10, 2019 (left) and March 8, 2024 (right). The lower photos taken April 20, 2012 (left) and April 22, 2022 (right).

One of my favorite case studies is of an ancestor coast live oak near Loma Mar, CA. This oak could not even be easily seen when I first arrived due to the dense growth of young Douglas firs and bay laurels. After removing the young trees that were crowding the oak and treating it with soil amendments and limewash the canopy of the oak has improved markedly in just five years (*Figure* 58).

Figure 58. An ancestor oak in Loma Mar, CA, where we removed encroaching Douglas firs and bays (upper photo set) followed by several soil fertilization and limewash treatments. Notice how the oak has responded with a fuller, lusher canopy (lower photo sets). Upper photo set taken a day apart, March 3, 2017 (left) and March 4, 2017 (right). Middle and lower photos taken five years apart, March 3, 2017 (left) and March 3, 2022 (right).

CHAPTER 9

Most of the before-and-after photos shown here were selected because of the nearness in date, time of day, and light condition of the two images. However, they also do tend to emphasize the more impressive positive responses, and are not representative of all my results. Indeed, some oaks do not respond well to treatments, and the health of a few appeared to worsen afterwards. In a 2008 paper, through analysis of more than 150 sets of before-and-after photos, I estimated that about 20 percent of the oaks I have treated with fire mimicry show either no change or a decrease in canopy density in following years.[2]

On the other hand, what happens when the oaks respond so well that a client decides, despite my recommendations, to opt out of further treatments? This occurs on occasion and the results are as might be expected. *Figure* 59 shows how the health of coast live oaks, initially on the mend, may lapse without further treatments.

Figure 59. Three coast live oaks in Piedmont, CA responding to fire mimicry treatments, but then declining after a lapse of treatments. Photo dates March 30. 2012 (left), April 1, 2013 (middle), and April 20, 2016 (right).

I admit that these results are rather wanting in control trees, but this is due to the fact that my clients usually prefer not to leave any of their trees untreated. Still, I have documented a handful of cases with control trees (those that were not treated with fire mimicry growing adjacent to those that were treated). These indicate that the treated coast live oaks responded well, whereas nearby untreated oaks did not (*Figure* 60).

Figure 60. Before-and-after photos of two coast live oaks in Atherton, CA taken on December 3, 2018 (left) and December 5, 2019 (right). The upper set show a treated oak, the lower set are of a nearby untreated oak.

Results following wildfires

In some cases I have treated oaks with fire mimicry only to see these trees burned in subsequent wildfires. These cases, while uncommon, are particularly instructive as to the efficacy of fire mimicry in helping trees survive wildfires.

The first case study involves a property in Sonoma, CA, that was affected by a large wildfire in 2017. In 2009, I began fire mimicry treatments on about five acres of oak woodland there. I spent several weeks clearing, thinning, and pruning, followed by pile-burning of the cut materials. I then applied organic compost, compost tea, and mineral amendments to the soils, along with removing moss and limewashing of the trunks. The soil amendments and limewashing were repeated every other year.

Photos taken soon after the 2017 wildfire show a high severity of fire damage to the adjacent, unmanaged forest, but very little damage to the trees on the treated property (*Figure* 61). In addition, even though the fire burned to within twenty feet of the home, the home suffered no structural damage, while less than ten trees out of hundreds on the property were killed. In this case I believe the work we did not only saved the home, it also saved the forest.

CHAPTER 9

Figure 61. A mixed oak forest in Sonoma, CA, treated with fire mimicry beginning in September of 2009. Approximately five acres of forest were cleared, pruned, and thinned, followed by the pile-burning of debris. Most of the oaks also received soil fertilization and limewash treatments prior to the 2017 wildfire. The upper photo was taken on November 14, 2017, soon after the Sonoma Fire, and show treated forests on the left, and untreated forests on the right. Note the fence line, where fire treatments stopped, visible in the right half of the photo. The lower photo set shows the site of pile-burning in on December 13, 2012 (left), before and November14, 2017 (right) after the 2017 Sonoma Fire. All the oaks visible in the background of the lower photos survived the fire, which thankfully stayed on the ground over most of the property.

The second case study involves a wildfire that burned through several groves of ancestor oaks previously treated with fire mimicry here in Big Sur. In 2019, I implemented fire mimicry treatments on a property with several ancient coast live oaks. This involved clearing brush, thinning young trees, pruning mature trees, amending the soils, and limewashing 21 oaks. A year later, I inspected the trees and found the health of most of had improved. However, in 2020, the Dolan Fire[3] burned through the property and destroyed several structures and large oaks. Of the 21 trees initially treated in 2019, seventeen survived the fire, and several even showed improvement a year or so after the fire (*Figure* 62). In hindsight, while we may not have done enough fuel reduction to save some of the structures on this property, our work did allow most of the old-growth oak groves to survive.

Figure 62. Coast live oaks in Big Sur, CA previously treated with fire mimicry that were affected by the Dolan Fire in 2020. These photo sets show several oaks at the time of treatment on February 25, 2019 (left), one year later on February 28, 2020, just prior to the Dolan Fire (center), and a year and a half after the Dolan Fire on February 24, 2022 (right).

My fire mimicry work has mainly involved coast live oaks, which seem to be the trees in most need of care here on the Central Coast of California. However, I have applied fire mimicry to several other tree species fairly good success. Positive results have been seen with valley oaks, Pacific madrones, Monterey pines, ponderosa pines, California buckeyes, coast redwoods, big leaf maples, and most fruit trees.[4]

Results of canker surgeries

Finally, I want to present a set of results that I'm especially pleased with — the response of the oaks to canker surgeries. After conducting hundreds of surgeries on bleeding stem cankers in oaks, many of which are likely due to Sudden Oak Death disease, I've found that most of the surgeries are successful, as indicated by the lack of any residual bleeding at or around the surgical wounds after multiple years. I've also noticed that in a majority of cases, the healing of the wounds, as seen in callus formation along the wound margin, is apparent in as little as one year.

One of the most dramatic responses I've seen to a canker surgery involves a coast live oak in Los Altos, CA. In 2017 I noticed two bleeding stem canker infections (possibly Sudden Oak Death) in the trunk of the oak. Following the removal of the cankers and mineral/compost tea fertilization of the soils, I have returned yearly to document the healing of the surgery. *Figure 63* presents a time series of photographs I've taken showing rapid healing of the surgical wounds that have now closed after five years with no residual infection! For some silly reason I do not have photos showing the change in the canopy health, but let me assure readers this oak is thriving.

Figure 63. Rapid healing of two surgical wounds in a coast live oak. Note there is no sign of residual infection. Top row (left to right): pre surgery, one year post-surgery, two years post-surgery. Bottom row (left to right): three years post-surgery, four years post-surgery, five years post-surgery.

Another example of a rapid recovery from a stem canker surgery (plus soil fertilization and limewash treatments) concerns a coast live oak in Woodside, CA. This tree developed a small stem canker that I removed in 2012, and I have since periodically documented the mending of the wound. *Figure* 64 shows the progression of wound healing after four years, six years, eight years, and eleven years. By year eleven the wound has completely healed closed with no sign of residual infection. As was hoped, the canopy of this oak also displayed marked improvement after the surgery and several fire mimicry treatments.

CHAPTER 9

Figure 64. Healing of a surgical wound in a coast live oak over nine years. Upper left, four years post-surgery, upper right, six years post-surgery, middle left, eight years post-surgery, and middle right, eleven years post-surgery. Note there is no residual sign of infection. The lower photos of the canopy of the same tree were taken (prior to surgery) on September 18, 2010 (left), and (seven years post-surgery) on October 15, 2019 (right).

After performing surgeries on some very large cankers that failed miserably, I have also learned there is a limit to the usefulness of this technique on severely diseased trees. Still, I am having some success, so far, in seeing wounds recover after larger surgeries.

I have also tried surgical procedures on other oaks, including valley oaks. While valley oaks only rarely get stem canker infections, I have found them in a few trees and successfully removed them.

Whether or not individual surgeries have ended up being effective, they have taught me many things about stem canker diseases in oaks. First, it is critically important to remove *all* infected (discolored) tissue so that only wood and healthy (reddish) bark remains. Second, cauterization is, I believe, the crucial stimulus needed for rapid callus formation and healing of the wound. Cauterization may also help to sterilize any lingering infected tissue and create a charred, antiseptic surface that can inhibit secondary infections. Third, during my surgeries I nearly always encounter multiple insect pests and pathogens inhabiting the infected tissues. Along with the *Phytophthera ssp.* and *Armillaria sp.* disease cankers, I find a range of insect pests such as oak borer larva, pill bugs, termites, and carpenter ants. When tracking the spread of infection during surgeries, it is further obvious that the presence of diseased tissue is closely associated with areas of high insect activity, both past and present.

While surgical intervention to remove disease is necessary at times, it must also be accompanied by other fire mimicry treatments. It would make little sense to do a surgery without addressing the root causes of the canker, which often lie in forest overcrowding, moss and lichen buildup, and/or poor fertility of the soils due to systemic acidification. The successful application of surgery to control diseases and pests, combined with the above stated methods of tree and soil care, demonstrates how Western surgical intervention can complement more traditional healing practices.

User-friendly fire mimicry

That, my friends, is a summary of the fire mimicry practices I have been employing for the past twenty years, along with my results. If these procedures appear simple, that is by design. In using commonly available materials and tools that are low cost, this basic protocol, aside from the surgical procedures, can be implemented by nearly anyone, anywhere. Readers may have noticed that there are no synthetic chemicals used in this protocol. The materials are all natural, nontoxic compounds typically used in organic agriculture. Following up on the work with before-and-after canopy photos of treated (and untreated) trees is easy and rewarding.

While I believe my methods are inherently sound, I feel there is plenty of room for improvement and I trust that these methods will continue to evolve for the better. In the following Epilogue, I will share my thoughts on how and where the practice of fire mimicry can aid us in the future.

EPILOGUE
REMEMBERING OUR PLACE IN NATURE

Through the course of researching and writing this book I have come to learn that a common trait among most of the world's Indigenous Peoples, both past and present, is the cultural practice of tending trees. Too extensive to have discussed here are volumes of ethnographic studies and oral histories that show how Indigenous Peoples, from the tropics to the subarctic and subantarctic, thrived and are still thriving among forests they have cultivated for millennia. Through reciprocal acts of tending the plants and soils, and receiving their sustenance, the People have learned how to best fill their various niches in the forest and help maintain a long-term relationship. That is why, today, we find the vast majority of the world's biodiversity hotspots on Indigenous lands.[1]

In every person's ancestry there are forest dwelling and tree tending Indigenous Peoples whose ecological knowledge must still linger in our DNA. Once this acumen is reawakened, we can then remember our place in nature and undertake informed actions that are guided by our ancestors' deep connections with the trees and forests.

EPILOGUE

My hope is that the knowledge shared here will inspire others to form a closer kinship with their surrounding trees and forests. I know that education is key and have already instructed hundreds of colleagues and students on these techniques, collaborating with the Mutsun Costanoan and Esselen Tribes in presenting Traditional Ecological Knowledge during cultural fire and fire mimicry training events on Esselen lands, and at Indian Canyon[2]. Several of these workshops have been in partnership with EcoCamp Coyote, Santa Cruz Permaculture, and the Central Coast Prescribed Burn Association.[3] Many of the participants have gone on to be regular practitioners of fire mimicry and cultural burning, and are already admiring their own results.

Furthermore, I am hopeful that this book will serve to underscore and broaden the understanding of the benefits of cultural fire (*Figure* 65). My good friend David Shearer, after reviewing it, commented "You are making the case that we need to rebrand fire as a solution strategy (good fire) versus a fundamental element to be fearful of (bad fire)." Precisely! And might I add that when "bad fires" (highly destructive canopy fires) do occur, they should be seen as windows of opportunity to alter the management trajectory of these burned-down forests. For all the damage caused by catastrophic wildfires, they also accomplish an immense amount of clearing, thinning, and pruning work that could never be done by hand at that scale. Rather than leaving these areas unmanaged to recover on their own, cultural fires, at some appropriate frequency, should be incorporated into a new management plan.

Catastrophic wildfires are just one among many forces threatening our forests, which include climate change, disease outbreaks, and insect infestations. Looking beyond these threats, Daniel Wildcat, in his book *On Indigenuity: Learning the Lessons of Mother Earth,* offers hope from an Indigenous perspective: "My hope resides in the belief that a fair amount of humankind's bad behavior is an artifact of the cultural lens through which most modern humankind understands the biosphere and our — dare I say — natural role in it."[4]

In my opinion any endeavor to restore our trees and forests will have the best chance of success when it takes place under the advice and guidance of the local Native Peoples.[5] I think I have already made a strong case for the wisdom of this approach. Had local land managers acted upon the Traditional Ecological Knowledge shared with them by Fred Nason and other members of the Esselen tribe in the 1970s, who knows, I may not have lost my home in Big Sur forty years later!

Figure 65. Ancestor oak canopies being "saged" by a cultural burn on Esselen land.

It is a given, in my world, that restoring and maintaining healthy forests at a landscape scale will require a lot more prescribed burning and cultural fires. Thankfully, more and more fire bosses and prescribed fire specialists are collaborating with culturally informed Tribal Elders such as Tom Little Bear Nason of the Esselen Tribe, Ron Goode of the North Fork Mono Tribe, Bill Tripp and Frank Lake of the Karuk Tribe, and Ali Meders-Knight of the Mechoopda Indian Tribe of Chico Rancheria.[6]

Bill Tripp has recently been appointed by President Biden to the Wildland Fire Mitigation and Management Commission a federal effort to help mitigate catastrophic wildfires. The Yurok Tribe has established a Cultural Fire Management Council whose intention is to "facilitate the practice of cultural burning on the Yurok Reservation and Ancestral lands."[7] And the Esselen, Karuk, Yurok, and North Fork Mono Tribes have all taken leadership roles in the education and training in the use of cultural fire to wisely manage native forests and grasslands.[8]

I also think there is a calling to Western science to help fill some of the gaps in our understanding of cultural forest tending and fire mimicry. If I had a wish list of scientific studies that could be undertaken to fill these gaps it would include, in no particular order:

- Perform more detailed, controlled studies on the efficacy of limewash on tree health.
- Carefully examine the ecological impacts, whether beneficial, benign, or harmful, of mosses and lichens on trees and soils. (This work is long overdue!)
- Utilize improved methods for the documentation and age determination of ancestor trees.
- Expand testing on the practicality and effectiveness of fire mimicry treatments for various tree species, groves/orchards of trees, and even entire watersheds.
- Develop technology to improve the photo-documentation of results using both ground-based and aerial imaging. Daily time-lapse photos of a tended grove of oaks adjacent to an untended grove, over

several years, would provide valuable evidence regarding the efficacy of fire mimicry.

I trust there are a number of inquisitive scientists who will eventually step up and help address some of these research questions. Each of these topics has been given fair attention in my own research and practice, so part of the scientific groundwork is already laid.

In the end, the vision of a healthy forest culture can only be realized by generations of people caring for the land, the trees, and each other through frequent acts of reciprocity, according to ethics that have long been central to Native traditions. It's not hard for me to see such a future, especially given that we have a trove of Indigenous Knowledge, now supported by a wide range of modern technologies.

I believe in this vision mainly because I know, even without modern technologies, that it has already been achieved by the Indigenous Peoples of California, throughout the Americas, and elsewhere around the world, for more than ten thousand years![9]

APPENDIX
SPECIES NOMENCLATURE OF TREES AND SHRUBS

Common Name - *Latin Name*
apple (Golden Delicious) - *Malus domestica*
aspen, quaking - *Populus tremuloides*
bay laurel, California - *Umbellularia californica*
beech, southern - *Nothofagus* sp.
birch, downy - *Betula pubescens*
buckeye, California - *Aesculus californica*
cedar, incense - *Calocedrus decurrans*
cypress, Alaska yellow - *Chamaecyparis nootkatensis*
cypress, Monterey - *Cupressus macrocarpa*
cypress, white - *Callitris columellaris*
elderberry - *Sambucus* sp.
fig, common - *Ficus carica*
fir, red - *Abies magnifica*
fir, Santa Lucia - *A. bracteata*
fir, subalpine - *A. lasiocarpa*
fir, Douglas - *Pseudotsuga menziesii*
gooseberry - *Ribes* sp.
gum, red - *Eucalyptus camaldulensis*

hazelnut - *Corylus cornuta*
hemlock, eastern - *Thuja canadensis*
hemlock, mountain - *T. mertensiana*
hemlock, western - *T. heterophylla*
juniper, California - *Juniperus californica*
larch, eastern - *Larix laricina*
madrone, Pacific - *Arbutus menziesii*
maple, bigleaf - *Acer macrophyllum*
maple, sugar - *A. saccharum*
oak, black - *Quercus kelloggii*
oak, blue - *Q. douglasii*
oak, canyon live - *Q. chrysolepis*
oak, coast live - *Q. agrifolia*
oak, Engelmann - *Q. engelmannii*
oak, English - *Q. robur*
oak, holm - *Q. rotundifolia*
oak, interior live - *Q. wislizeni*
oak, Oregon white - *Q. garryana*
oak, sessile - *Q. petraea*
oak, southern live - *Q. virginiana*
oak, Spanish - *Q. pyrenaica*
oak, valley - *Q. lobata*
oak, tanbark - *Lithocarpus densiflorus*
pine, gray - *Pinus sabiniana*
pine, Jeffrey - *P. jeffreyi*
pine, lodgepole - *P. contorta*
pine, Masson - *P. massoniana*
pine, Monterey - *P. radiata*
pine, ponderosa - *P. ponderosa*
pine, Scots - *P. sylvestris*
poison-oak, Pacific - *Toxicodendron diversilobum*
redcedar, western - *Thuja plicata*
redwood, coast - *Sequoia sempervirens*
sage, black - *Salvia* sp.

APPENDIX

sage, white - *Artemesia* sp.
sequoia, giant - *Sequoiadendron giganteum*
silk tassel, coastal - *Garrya elliptica*
spruce, black - *Picea mariana*
spruce, Engelmann - *P. engelmannii*
spruce, Norway - *P. abies*
spruce, red - *P. rubens*
spruce, Sitka - *P. sitchensis*
sycamore, California - *Platanus racemosa*
toyon - *Heteromeles arbutifolia*

BIBLIOGRAPHY

Agbeshie, A.A., Abugre, S., Atta-Darkwa, T., Awuah, R., 2022. "A review of the effects of forest fire on soil properties." *Journal of Forestry Research* 33, 1419–1441. https://doi.org/10.1007/s11676-022-01475-4

Aldern, J.D., Goode, R.W., 2014. "The stories hold water: Learning and burning in North Fork Mono homelands." *Decolonization: Indigeneity, Education, and Society* 3, 26–51.

Anderson, E., 1952. *Plants, Man and Life*. Boston: Little, Brown & Co.

Anderson, L., Carlson, C.E., Wakimoto, R.H., 1987. "Forest fire frequency and western spruce budworm outbreaks in western Montana." *Forest Ecology and Management* 22, 251–260. https://doi.org/10.1016/0378-1127(87)90109-5

Anderson, M.K., 2006. "The use of fire by Native Americans in California," in: Agee, J.K. (Ed.), *Fire in California's Ecosystems*. Berkeley: University of California Press, pp. 417–430.

Anderson, M.K., 2005. *Tending the Wild: Native American Knowledge and the Management of California's Natural Resources*. Berkeley: University of California Press.

Anderson, M.K., 1994. "Prehistoric anthropogenic wildland burning by hunter-gatherer societies in the temperate regions: A net source, sink, or neutral to the global carbon budget?" *Chemosphere* 29, 913–934.

Anderson, M.K., Stewart, O.C., 2002. "An ecological critique," in: *Forgotten Fires: Native Americans and the Transient Wilderness*. Norman: University of Oklahoma Press, pp. 37–64.

Armstrong, C.G., Earnshaw, J., McAlvay, A.C., 2022. "Coupled archaeological and ecological analyses reveal ancient cultivation and land use in Nuchatlaht (Nuu-chah-nulth) territories, Pacific Northwest." *Journal of Archaeological Science* 143, 105611. https://doi.org/10.1016/j.jas.2022.105611

Arneth, A., Niinemets, Ü., Pressley, S., Bäck, J., Hari, P., Karl, T., Noe, S., Prentice, I.C., Serça, D., Hickler, T., Wolf, A., Smith, B., 2007. "Process-based estimates of terrestrial ecosystem isoprene emissions: incorporating the effects of a direct CO-isoprene interaction." *Atmospheric Chemistry and Physics* 7, 31–53. https://doi.org/10.5194/acp-7-31-2007

Arno, S.F., Fiedler, C.E., 2005. *Mimicking Nature's Fire*. Washington DC: Island Press.

Ascaso, C., Rapsch, S., 1986. "Ultrastructural changes in chloroplasts of *Quercus rotundifolia* Lam. in response to evernic acid." *Annals of Botany* 57, 407–413. https://doi.org/10.1093/oxfordjournals.aob.a087123

Asouti, E., Kabukcu, C., 2014. "Holocene semi-arid oak woodlands in the Irano-

BIBLIOGRAPHY

Anatolian region of Southwest Asia: natural or anthropogenic?" *Quaternary Science Reviews* 90, 158–182. https://doi.org/10.1016/j.quascirev.2014.03.001

Axelrod, D.I., 1982. "Age and origin of the Monterey endemic area." *Madroño* 29, 127–147.

Bakker, M.R., Kerisit, R., Verbist, K., Nys, C., 2000. "Effects of liming on rhizosphere chemistry and growth of fine roots and of shoots of sessile oak (*Quercus petraea*)," in: Stokes, A. (Ed.), *The Supporting Roots of Trees and Woody Plants: Form, Function and Physiology, Developments in Plant and Soil Sciences.* Dordrecht: Springer Netherlands, pp. 405–417. https://doi.org/10.1007/978-94-017-3469-1_40

Bakker, M.R., Nys, C., 1999. "Effects of lime-induced differences in site fertility on fine roots of oak." *Annals of Forest Science* 56, 599–606. https://doi.org/10.1051/forest:19990707

Baldwin, I.T., Halitschke, R., Paschold, A., Von Dahl, C.C., Preston, C.A., 2006. "Volatile signaling in plant-plant interactions: 'Talking Trees' in the genomics era." *Science* 311, 812–815. https://doi.org/10.1126/science.1118446

Barrett, S.W., Arno, S.F., 1999. "Indian fires in the Northern Rockies," in: Boyd, R. (Ed.), *Indians Fire and the Land in the Pacific Northwest*. Corvallis: Oregon State University Press, pp. 50–64.

Battles, J.J., Fahey, T.J., Driscoll, C.T., Blum, J.D., Johnson, C.E., 2014. "Restoring soil calcium reverses forest decline." *Environmental Science & Technology Letters* 1, 15–19. https://doi.org/10.1021/ez400033d

Beh, M., Metz, M., Frangioso, K., Rizzo, D., 2012. "Survival of *Phytophthora ramorum* following wildfires in the Sudden Oak Death-impacted forests of the Big Sur region." *Proceedings of the Sudden Oak Death Fifth Science Symposium USDA General Technical Report* PSW-GTR-243, 62–64.

Biederman, L.A., Harpole, W.S., 2013. "Biochar and its effects on plant productivity and nutrient cycling: a meta-analysis." *GCB Bioenergy* 5, 202–214. https://doi.org/10.1111/gcbb.12037

Bowcutt, F., 2013. "Tanoak landscapes: Tending a Native American nut tree." *Madroño* 60, 64–86. https://doi.org/10.3120/0024-9637-60.2.64

Brennan, R.N., Boychuck, S., Washkwich, A.J., John-Alder, H., Fonseca, D.M., 2023. "Tick abundance and diversity are substantially lower in thinned vs. unthinned forests in the New Jersey Pinelands National Reserve, USA." *Ticks and Tick-borne Diseases* 14, 102106. https://doi.org/10.1016/j.ttbdis.2022.102106

Breschini, G.S., Haversat, T., 2004. *The Esselen Indians of the Big Sur Country*. Salinas: Coyote Press.

Brown, P.M., Baxter, W.T., 2003. "Fire history in coast redwood forests of the Mendocino coast, California." *Northwest Science* 77, 147–158.

Cajete, G., 2006. "Western science and the loss of natural creativity," in: Four Arrows Don Trent Jacobs (Ed.), *Unlearning the Language of Conquest: Scholars Expose Anti-Indianism in America*. Austin: University of Texas Press, pp. 247–259.

Chiapusio, G., Jassey, V.E.J., Hussain, M.I., Binet, P., 2013. "Evidences of bryophyte allelochemical interactions: The case of *Sphagnum*," in: Cheema, Z.A., Farooq,

BIBLIOGRAPHY

M., Wahid, A. (Eds.), *Allelopathy*. Heidelberg: Springer, pp. 39–54. https://doi.org/10.1007/978-3-642-30595-5_3

Clements, F.E., 1916. *Plant Succession: An Analysis of the Development of Vegetation*. Washington DC: Carnegie Institution of Washington.

Close, D.C., Davidson, N.J., Johnson, D.W., Abrams, M.D., Hart, S.C., Lunt, I.D., Archibald, R.D., Horton, B., Adams, M.A., 2009. "Premature decline of eucalyptus and altered ecosystem processes in the absence of fire in some Australian forests." *The Botanical Review* 75, 191–202. https://doi.org/10.1007/s12229-009-9027-y

Cornish, M., 1999. "Forest decline as a successional process: The role of bryophytes in a montane ecosystem in the Colorado Rocky Mountains" (M.Sc. diss.), University of Oxford, Oxford U.K.

Crawford, J.N., Mensing, S.A., Lake, F.K., Zimmerman, S.R., 2015. "Late Holocene fire and vegetation reconstruction from the western Klamath Mountains, California, USA: A multi-disciplinary approach for examining potential human land-use impacts." *The Holocene* 25, 1341–1357. https://doi.org/10.1177/0959683615584205

Davis, K.T., Peeler, J., Fargione, J., Haugo, R.D., Metlen, K.L., Robles, M.D., Woolley, T., 2024. "Tamm review: A meta-analysis of thinning, prescribed fire, and wildfire effects on subsequent wildfire severity in conifer dominated forests of the Western US." *Forest Ecology and Management* 561, 121885. https://doi.org/10.1016/j.foreco.2024.121885

Delmas, R.A., Druilhet, A., Cros, B., Durand, P., Delon, C., Lacaux, J.P., Brustet, J.M., Serça, D., Affre, C., Guenther, A., Greenberg, J., Baugh, W., Harley, P., Klinger, L., Ginoux, P., Brasseur, G., Zimmerman, P.R., Grégoire, J.M., Janodet, E., Tournier, A., Perros, P., Marion, Th., Gaudichet, A., Cachier, H., Ruellan, S., Masclet, P., Cautenet, S., Poulet, D., Biona, C.B., Nganga, D., Tathy, J.P., Minga, A., Loemba-Ndembi, J., Ceccato, P., 1999. "Experiment for Regional Sources and Sinks of Oxidants (EXPRESSO): An overview." *Journal of Geophysical Research: Atmospheres* 104, 30609–30624. https://doi.org/10.1029/1999JD900291

Doolittle, W.E., 2000. *Cultivated Landscapes of Native North America*. Oxford: Oxford University Press.

Douhovnikoff, V., Cheng, A.M., Dodd, R.S., 2004. "Incidence, size and spatial structure of clones in second-growth stands of coast redwood, *Sequoia sempervirens* (Cupressaceae)." *American Journal of Botany* 91, 1140–1146. https://doi.org/10.3732/ajb.91.7.1140

Downie, A.E., Van Zwieten, L., Smernik, R.J., Morris, S., Munroe, P.R., 2011. "Terra Preta Australis: Reassessing the carbon storage capacity of temperate soils." *Agriculture, Ecosystems & Environment* 140, 137–147. https://doi.org/10.1016/j.agee.2010.11.020

Drury, A. (Ed.), 1954. *Point Lobos Reserve: Interpretation of a Primitive Landscape*. State of California: Department of Natural Resources.

Drury, A., Neasham, V.A., 1954. "History of Point Lobos," in: Drury, A. (Ed.), *Point

BIBLIOGRAPHY

Lobos Reserve: Interpretation of a Primitive Landscape. State of California: Department of Natural Resources, pp. 78–85.

Eicher, G.J., Rounsefell, G.A., 1957. "Effects of lake fertilization by volcanic activity on abundance of salmon." *Limnology & Oceanography* 2, 70–76. https://doi.org/10.4319/lo.1957.2.2.0070

Ellis, E.C., Gauthier, N., Klein Goldewijk, K., Bliege Bird, R., Boivin, N., Díaz, S., Fuller, D.Q., Gill, J.L., Kaplan, J.O., Kingston, N., Locke, H., McMichael, C.N.H., Ranco, D., Rick, T.C., Shaw, M.R., Stephens, L., Svenning, J.-C., Watson, J.E.M., 2021. "People have shaped most of terrestrial nature for at least 12,000 years." *Proceedings of the National Academy of Sciences* 118, e2023483118. https://doi.org/10.1073/pnas.2023483118

Eudoxie, G., Martin, M., 2019. "Compost tea quality and fertility," in: Larramendy, M., Soloneski, S. (Eds.), *Organic Fertilizers — History, Production and Applications.* London: IntechOpen, pp. 1–25. https://doi.org/10.5772/intechopen.86877

Fairhead, J., Fraser, J., Amanor, K., Solomon, D., Lehmann, J., Leach, M., 2017. "Indigenous soil enrichment for food security and climate change in Africa and Asia: A review," in: Stillitoe, P. (Ed.), *Indigenous Knowledge: Enhancing Its Contribution to Natural Resources Management.* Boston: CABI, pp. 99–115.

Fay, N., undated. "Retrenchment pruning and conservation arboriculture: Learning from old trees to develop natural management techniques." *Treeworks Environmental Practice* 1–18.

Fenn, M.E., Huntington, T.G., Mclaughlin, S.B., Eagar, C., Gomez, A., Cook, R.B., 2006. "Status of soil acidification in North America." *Journal of Forest Science* 52, S3–S13. https://doi.org/10.17221/10152-JFS

Fisher, J.A., Shackelford, N., Hocking, M.D., Trant, A.J., Starzomski, B.M., 2019. "Indigenous peoples' habitation history drives present-day forest biodiversity in British Columbia's coastal temperate rainforest." *People and Nature* 1, 103–114. https://doi.org/10.1002/pan3.16

Frantom, M., (undated). "Limewash: An old practice and a good one." US National Park Service. URL https://www.nps.gov/articles/limewash-an-old-practice-and-a-good-one.htm (accessed 5.10.23).

Fritz, Ö., Brunet, J., 2010. "Epiphytic bryophytes and lichens in Swedish beech forests — effects of forest history and habitat quality." *Ecological Bulletins* 53, 95–107.

Froelich, R.C., Hodges, C.S., Jr., Sackett, S.S., 1978. "Prescribed burning reduces severity of *Annosus* root rot in the South." *Forest Science* 24, 93–100. https://doi.org/10.1093/forestscience/24.1.93

Gallagher, M.R., Kreye, J.K., Machtinger, E.T., Everland, A., Schmidt, N., Skowronski, N.S., 2022. "Can restoration of fire-dependent ecosystems reduce ticks and tick-borne disease prevalence in the eastern United States?" *Ecological Applications* 32, e2637. https://doi.org/10.1002/eap.2637

Garrison, B.A., Otahal, C.D., Triggs, M.L., 2002. "Age structure and growth of California black oak (*Quercus kelloggii*) in the central Sierra Nevada, California."

BIBLIOGRAPHY

USDA Forest Service, Pacific Southwest Research Station, Redding, CA (PSW-GTR-184) 665–679.

Glime, J.M., 2024. "Roles of Bryophytes in Forest Sustainability—Positive or Negative?" *Sustainability* 16, 1–70. https://doi.org/10.3390/su16062359

Gordon, B.L., 1974. *Monterey Bay Area: Natural History and Cultural Imprints.* The Boxwood Press, Pacific Grove, CA.

Gossard, H.A., 1913. "Orchard bark beetles and pin hole borers." *Ohio Agricultural Experiment Station* 264, 1–68.

Greenlee, J.M., Langenheim, J.H., 1990. "Historic Fire Regimes and Their Relation to Vegetation Patterns in the Monterey Bay Area of California." *American Midland Naturalist* 124, 239–253. https://doi.org/10.2307/2426173

Greenler, S.M., Lake, F.K., Tripp, W., McCovey, K., Tripp, A., Hillman, L.G., Dunn, C.J., Prichard, S.J., Hessburg, P.F., Harling, W., Bailey, J.D., 2024. "Blending Indigenous and western science: Quantifying cultural burning impacts in Karuk Aboriginal Territory." *Ecological Applications* e2973. https://doi.org/10.1002/eap.2973

Griffin, J.R., Critchfield, W.B., 1976. "The distribution of forest trees in California." *USDA Forest Research Paper* PSW-82, 1-118.

Gusella, G., Morgan, D.P., Michailides, T.J., 2021. "Further investigation on limb dieback of fig (*Ficus carica*) caused by *Neoscytalidium dimidiatum* in California." *Plant Disease* 105, 324–330. https://doi.org/10.1094/PDIS-06-20-1226-RE

Haider, F.U., Wang, X., Zulfiqar, U., Farooq, M., Hussain, S., Mehmood, T., Naveed, M., Li, Y., Liqun, C., Saeed, Q., Ahmad, I., Mustafa, A., 2022. "Biochar application for remediation of organic toxic pollutants in contaminated soils; An update." *Ecotoxicology and Environmental Safety* 248, 114322. https://doi.org/10.1016/j.ecoenv.2022.114322

Halpern, A.A., Sousa, W.P., Lake, F.K., Carlson, T.J., Paddock, W., 2022. "Prescribed fire reduces insect infestation in Karuk and Yurok acorn resource systems." *Forest Ecology and Management* 505, 119768. https://doi.org/10.1016/j.foreco.2021.119768

Hamburg, S.P., Yanai, R.D., Arthur, M.A., Blum, J.D., Siccama, T.G., 2003. "Biotic control of calcium cycling in northern hardwood forests: Acid rain and aging forests." *Ecosystems* 6, 399–406. https://doi.org/10.1007/s10021-002-0174-9

Hamlet, L.E., 2014. "Anthropic sediments on the Scottish North Atlantic seaboard: Nature, versatility and value of midden" (Ph.D. diss.), University of Stirling, Scotland.

Hanson, C., 2021. *Smokescreen: Debunking Wildfire Myths to Save Our Forests and Our Climate.* Lexington: University of Kentucky Press.

Harmon, M.E., Franklin, J.F., 1989. "Tree seedlings on logs in *Picea-Tsuga* forests of Oregon and Washington." *Ecology* 70, 48–59.

Harraz, F.M., Hammoda, H.M., El-Hawiet, A., Radwam, M.M., Wanas, A.S., Eid, A.M., ElSohly, M.A., 2020. "Chemical constituents, antibacterial and acetylcholine esterase inhibitory activity of *Cupressus macrocarpa* leaves." *Natural Product Research* 34, 816–822.

BIBLIOGRAPHY

Hartweg, K.T., 1847. "Journal of a mission to California in search of plants." *Journal of the Horticultural Society* 2, 187–191.

Hauck, M., 2003. "Epiphytic lichen diversity and forest dieback: The role of chemical site factors." *The Bryologist* 106, 257–269.

Hawley, G.J., Schaberg, P.G., Eagar, C., Borer, C.H., 2006. "Calcium addition at the Hubbard Brook Experimental Forest reduced winter injury to red spruce in a high-injury year." *Canadian Journal of Forest Research* 36, 2544–2549. https://doi.org/10.1139/x06-221

Helbig, M., Waddington, J.M., Alekseychik, P., Amiro, B., Aurela, M., Barr, A.G., Black, T.A., Carey, S.K., Chen, J., Chi, J., Desai, A.R., Dunn, A., Euskirchen, E.S., Flanagan, L.B., Friborg, T., Garneau, M., Grelle, A., Harder, S., Heliasz, M., Humphreys, E.R., Ikawa, H., Isabelle, P.-E., Iwata, H., Jassal, R., Korkiakoski, M., Kurbatova, J., Kutzbach, L., Lapshina, E., Lindroth, A., Löfvenius, M.O., Lohila, A., Mammarella, I., Marsh, P., Moore, P.A., Maximov, T., Nadeau, D.F., Nicholls, E.M., Nilsson, M.B., Ohta, T., Peichl, M., Petrone, R.M., Prokushkin, A., Quinton, W.L., Roulet, N., Runkle, B.R.K., Sonnentag, O., Strachan, I.B., Taillardat, P., Tuittila, E.-S., Tuovinen, J.-P., Turner, J., Ueyama, M., Varlagin, A., Vesala, T., Wilmking, M., Zyrianov, V., Schulze, C., 2020. "The biophysical climate mitigation potential of boreal peatlands during the growing season." *Environmental Research Letters* 15, 104004. https://doi.org/10.1088/1748-9326/abab34

Helmig, D., Klinger, L.F., Guenther, A., Vierling, L., Geron, C., Zimmerman, P., 1999a. "Biogenic volatile organic compound emissions (BVOCs) I. Identifications from three continental sites in the U.S." *Chemosphere* 38, 2163–2187. https://doi.org/10.1016/S0045-6535(98)00425-1

Helmig, D., Klinger, L.F., Guenther, A., Vierling, L., Geron, C., Zimmerman, P., 1999b. "Biogenic volatile organic compound emissions (BVOCs) II. Landscape flux potentials from three continental sites in the U.S." *Chemosphere* 38, 2189–2204. https://doi.org/10.1016/S0045-6535(98)00424-X

Henrikson, M.I., 2017. *The Mystery of the Fire Trees of Southeast Alaska*. Ward Cove: Seventh Generation Arts.

Heusser, L., 1998. "Direct correlation of millennial-scale changes in western North American vegetation and climate with changes in the California Current System over the past ~60 kyr." *Paleoceanography* 13, 252–262. https://doi.org/10.1029/98PA00670

Hipp, A.L., Manos, P.S., González-Rodríguez, A., Hahn, M., Kaproth, M., McVay, J.D., Avalos, S.V., Cavender-Bares, J., 2018. "Sympatric parallel diversification of major oak clades in the Americas and the origins of Mexican species diversity." *New Phytologist* 217, 439–452. https://doi.org/10.1111/nph.14773

Hoffman, K.M., Gavin, D.G., Lertzman, K.P., Smith, D.J., Starzomski, B.M., 2016. "13,000 years of fire history derived from soil charcoal in a British Columbia coastal temperate rain forest." *Ecosphere* 7, e01415. https://doi.org/10.1002/ecs2.1415

Hoffman, K.M., Lertzman, K.P., Starzomski, B.M., 2017. "Ecological legacies of

BIBLIOGRAPHY

anthropogenic burning in a British Columbia coastal temperate rain forest." *Journal of Biogeography* 44, 2903–2915. https://doi.org/10.1111/jbi.13096

Hood, S.M., Baker, S., Sala, A., 2016. "Fortifying the forest: thinning and burning increase resistance to a bark beetle outbreak and promote forest resilience." *Ecological Applications* 26, 1984–2000. https://doi.org/10.1002/eap.1363

Huggett, B.A., Schaberg, P.G., Hawley, G.J., Eagar, C., 2007. "Long-term calcium addition increases growth release, wound closure, and health of sugar maple (*Acer saccharum*) trees at the Hubbard Brook Experimental Forest." *Canadian Journal of Forest Research* 37, 1692–1700. https://doi.org/10.1139/X07-042

Huggett, R.J., 1998. "Soil chronosequences, soil development, and soil evolution: a critical review." *Catena* 32, 155–172. https://doi.org/10.1016/S0341-8162(98)00053-8

Huntington, T.G., Hooper, R.P., Johnson, C.E., Aulenbach, B.T., Cappellato, R., Blum, A.E., 2000. "Calcium depletion in a southeastern United States forest ecosystem." *Soil Science Society of America Journal* 64, 1845–1858. https://doi.org/10.2136/sssaj2000.6451845x

Jepson, W., 1954a. "A 'tree island' of Monterey cypress," in: Drury, A. (Ed.), *Point Lobos Reserve: Interpretation of a Primitive Landscape.* State of California: Department of Natural Resources, pp. 39–43.

Jepson, W., 1954b. "Appendix 1: Scientific notes on the Monterey cypress," in: Drury, A. (Ed.), *Point Lobos Reserve: Interpretation of a Primitive Landscape.* State of California: Department of Natural Resources, pp. 86–87.

Jepson, W., 1923. *The Trees of California*, 2nd ed. Berkeley: University of California Press.

Jerardino, A., 2016. "On the origins and significance of Pleistocene coastal resource use in southern Africa with particular reference to shellfish gathering." *Journal of Anthropological Archaeology* 41, 213–230. https://doi.org/10.1016/j.jaa.2016.01.001

Johnson, C.E., Driscoll, C.T., Blum, J.D., Fahey, T.J., Battles, J.J., 2014. "Soil chemical dynamics after calcium silicate addition to a northern hardwood forest." *Soil Science Society of America Journal* 78, 1458–1468. https://doi.org/10.2136/sssaj2014.03.0114

Johnson, D.L., 1977. "The late Quaternary climate of coastal California: Evidence for an ice age refugium." *Quaternary Research* 8, 154–179.

Jones, T.L., 2003. "Prehistoric Human Ecology of the Big Sur Coast," *Contributions of the University of California Archeological Research Facility* 61, 1-291.

Juice, S.M., Fahey, T.J., Siccama, T.G., Driscoll, C.T., Denny, E.G., Eagar, C., Cleavitt, N.L., Minocha, R., Richardson, A.D., 2006. "Response of sugar maple to calcium addition to northern hardwood forest." *Ecology* 87, 1267–1280. https://doi.org/10.1890/0012-9658(2006)87[1267:ROSMTC]2.0.CO;2

Jurskis, V., 2009. "River red gum and white cypress forests in south-western New South Wales, Australia: Ecological history and implications for conservation of grassy woodlands." *Forest Ecology and Management* 258, 2593-2601. https://doi.org/10.1016/j.foreco.2009.09.017

BIBLIOGRAPHY

Karst, J., Jones, M.D., Hoeksema, J.D., 2023. "Positive citation bias and overinterpreted results lead to misinformation on common mycorrhizal networks in forests." *Nature Ecology & Evolution* 7, 501–511. https://doi.org/10.1038/s41559-023-01986-1

Keeley, J.E., 1982. "Distribution of lightning- and man-caused wildfires in California." *General Technical Report Pacific Southwest Forest and Range Experiment Station* PSW-58, 431–437.

Kimmerer, R.W., 2014. "Returning the gift." *Minding Nature* 7, 18–24.

Kimmerer, R.W., 2013. *Braiding Sweetgrass: Indigenous Wisdom, Scientific Knowledge and the Teachings of Plants*. Minneapolis: Milkweed Editions.

Kimmerer, R.W., 2003. *Gathering Moss: A Natural and Cultural History of Mosses*. Corvallis: Oregon State University Press.

Kimmerer, R.W., Lake, F.K., 2001. "The role of indigenous burning in land management." *Journal of Forestry* 99, 36–41.

Klinger, L.F., 2009. "Forest vegetation and soil succession: The natural process of change." *Proceedings of the TEP Seminar XIII: Trees, Roots, Fungi, Soil* (Part 2), Treeworks Environmental Practice.

Klinger, L.F., 2008. "A holistic approach to mitigating pathogenic effects on trees." *Proceedings of the TEP Seminar XII: Trees, Roots, Fungi, Soil* (Part 1), Treeworks Environmental Practice.

Klinger, L.F., 2006. "Ecological evidence of large-scale silviculture by California Indians," in: Four Arrows Don Trent Jacobs (Ed.), *Unlearning the Language of Conquest: Scholars Expose Anti-Indianism in America*. Austin: University of Texas Press, pp. 153–165.

Klinger, L.F., 2005. "Bryophytes and soil acidification effects on trees: The case of Sudden Oak Death." *Combined Proceedings International Plant Propagators' Society* 55, 493–503.

Klinger, L.F., 2004. "Gaia and complexity," in: Schneider, S.H., Miller, J.R., Crist, E., Boston, P.J. (Eds.), *Scientists Debate Gaia: The Next Century*. Cambridge: The MIT Press, pp. 187–200.

Klinger, L.F., 1996. "The myth of the classic hydrosere model of bog succession." *Arctic and Alpine Research* 28, 1–9.

Klinger, L.F., 1992. "Peatland formation and ice ages: a possible Gaian mechanism related to vegetation succession," in: Schneider, S.H., Boston, P.J. (Eds.), *Scientists on Gaia*. Cambridge: The MIT Press, pp. 247–255.

Klinger, L.F., 1990. "Global patterns in community succession. 1. Bryophytes and forest decline." *Memoirs of the Torrey Botanical Club* 24, 1–50.

Klinger, L.F., 1988. "Successional change in vegetation and soils of Southeast Alaska" (Ph.D. diss.), University of Colorado, Boulder.

Klinger, L.F., Erickson, D.J., 1997. "Geophysiological coupling of marine and terrestrial ecosystems." *Journal of Geophysical Research: Atmospheres* 102, 25359–25370. https://doi.org/10.1029/97JD01620

Klinger, L.F., Short, S.K., 1996. "Succession in the Hudson Bay lowland, northern

BIBLIOGRAPHY

Ontario, Canada." *Arctic and Alpine Research* 28, 172–183. https://doi.org/10.2307/1551757

Klinger, L.F., Taylor, J.A., Franzen, L.G., 1996. "The potential role of peatland dynamics in ice-age initiation." *Quaternary Research* 45, 89–92. https://doi.org/10.1006/qres.1996.0008

Klinger, L.F., Walker, D.W., Webber, P.J., 1983. "The effects of gravel roads on Alaskan arctic coastal plain tundra," in: *Permafrost: Fourth International Conference, Proceedings. National Academy of Sciences*. Washington DC: National Academy Press, pp. 628–633.

Knight, C.A., Anderson, L., Bunting, M.J., Champagne, M., Clayburn, R.M., Crawford, J.N., Klimaszewski-Patterson, A., Knapp, E.E., Lake, F.K., Mensing, S.A., Wahl, D., Wanket, J., Watts-Tobin, A., Potts, M.D., Battles, J.J., 2022. "Land management explains major trends in forest structure and composition over the last millennium in California's Klamath Mountains." *Proceedings of the National Academy of Sciences* 119, e2116264119. https://doi.org/10.1073/pnas.2116264119

Koenig, W.D., Knops, J.M.H., 2005. "The mystery of masting in trees." *American Scientist* 93, 340–347.

Kroeber, A.L., 1925. *Handbook of the Indians of California*. New York: Dover Publications Inc.

Kroeber, A.L., 1907. *Indian Myths of South Central California*. Berkeley: University of California Publications.

La Perouse, J.F., 1807. *A Voyage Round the World* [excerpt from 1786]. London: Lackington, Allen, and Co.

Lake, F.K., Christianson, A.C., 2019. "Indigenous fire stewardship," in: Manzello, S.L. (Ed.), *Encyclopedia of Wildfires and Wildland-Urban Interface (WUI) Fires*. Cham: Springer International Publishing, pp. 1–9. https://doi.org/10.1007/978-3-319-51727-8_225-1

Lake, F.K., Wright, V., Morgan, P., McFadzen, M., McWethy, D., Stevens-Rumann, C., 2017. "Returning fire to the land: Celebrating Traditional Knowledge and fire." *Journal of Forestry* 115, 343–353. https://doi.org/10.5849/jof.2016-043R2

Langenheim, J.H., Durham, J.W., 1963. "Quaternary closed-cone pine flora from travertine near Little Sur, California." *Madroño* 17, 33–51.

Legaz, M.E., Monsó, M.A., Vicente, C., 2004. "Harmful effects of epiphytic lichens on trees." *Recent Research Developments in Agronomy and Horticulture* 1, 1–10.

Legaz, Me., Perez-Urria, E., Avalos, A., Vicente, C., 1988. "Epiphytic lichens inhibit the appearance of leaves in *Quercus pyrenaica*." *Biochemical Systematics and Ecology* 16, 253–259. https://doi.org/10.1016/0305-1978(88)90002-6

Leopold, A., 1972. *Round River*. Oxford: Oxford University Press.

Levy, S., 2022. "The return of intentional forest fires." *BioScience* 72, 324–330. https://doi.org/10.1093/biosci/biac016

Li, Y., Cui, S., Chang, S.X., Zhang, Q., 2019. "Liming effects on soil pH and crop yield depend on lime material type, application method and rate, and crop species: a global meta-analysis." *Journal of Soils and Sediments* 19, 1393–1406. https://doi.org/10.1007/s11368-018-2120-2

Li, Z., Wang, Y., Liu, Y., Guo, H., Li, T., Li, Z.-H., Shi, G., 2014. "Long-term effects of liming on health and growth of a Masson pine stand damaged by soil acidification in Chongqing, China." *PLoS ONE* 9, e94230. https://doi.org/10.1371/journal.pone.0094230

Lightfoot, K.G., Parrish, O., 2009. *California Indians and Their Environment: An Introduction*. Berkeley: University of California Press.

Long, J.N., 2009. "Emulating natural disturbance regimes as a basis for forest management: A North American view." *Forest Ecology and Management* 257, 1868–1873. https://doi.org/10.1016/j.foreco.2008.12.019

Long, J.W., Goode, R.W., Gutteriez, R.J., Lackey, J.J., Anderson, M.K., 2017. "Managing California black oak for Tribal ecocultural restoration." *Journal of Forestry* 115, 426–434. https://doi.org/10.5849/jof.16-033

Long, J.W., Lake, F.K., Goode, R.W., Burnette, B.M., 2020. "How Traditional Tribal perspectives influence ecosystem restoration." *Ecopsychology* 12, 71–82. https://doi.org/10.1089/eco.2019.0055

Lorimer, C.G., Porter, D.J., Madej, M.A., Stuart, J.D., Veirs, S.D., Norman, S.P., O'Hara, K.L., Libby, W.J., 2009. "Presettlement and modern disturbance regimes in coast redwood forests: Implications for the conservation of old-growth stands." *Forest Ecology and Management* 258, 1038–1054. https://doi.org/10.1016/j.foreco.2009.07.008

Lovelock, J., 1995. *The Ages of Gaia*, 2nd ed. London: W.W. Norton & Co.

Lutz, J.A., Van Wagtendonk, J.W., Thode, A.E., Miller, J.D., Franklin, J.F., 2009. "Climate, lightning ignitions, and fire severity in Yosemite National Park, California, USA." *International Journal of Wildland Fire* 18, 765. https://doi.org/10.1071/WF08117

MacBride, J.R., Froelich, D., 1984. "Structure and condition of older stands in parks and open spaces of San Francisco, California." *Urban Ecology* 8, 165–178.

MacBride, T.H., 1913. "The Monterey conifers." *Proceedings of the Iowa Academy of Science* 20, 1–7.

Maehr, T., 2024. *Fire on the Mountain - Living with Wildfire in the Santa Lucia Mountains of Big Sur* (in press).

Mann, C.C., 2011. *1491: New Revelations of the Americas Before Columbus*, 2nd ed. New York: Vintage Books.

Mansion, D., undated. "Trognes, Tétards, Emondes, Plesses: The multiple aspects and uses of the farmer's oak." www.internationaloaksociety.org.

Marianchild, K., 2014. *Secrets of the Oak Woodland: Plant and Animals among California Oaks*. Berkeley: Heyday.

Martinez, D., 2018. "Redefining sustainability through kincentric ecology: reclaiming indigenous lands, knowledge, and ethics," in: Nelson, M.K., Shilling, D. (Eds.), *Traditional Ecological Knowledge: Learning from Indigenous Practices for Environmental Sustainability*. Cambridge: Cambridge University Press, pp. 139–174.

Martinez, D.J., Cannon, C.E.B., McInturff, A., Alagona, P.S., Pellow, D.N., 2023.

BIBLIOGRAPHY

"Back to the future: Indigenous relationality, kincentricity and the North American Model of wildlife management." *Environmental Science and Policy* 140, 202–207. https://doi.org/10.1016/j.envsci.2022.12.010

Masters, R., 2015. "Summer solstice at a California stonehenge" [WWW Document]. *Hilltromper Santa Cruz*. URL https://hilltromper.com/article/summer-solstice-california-stonehenge (accessed 5.8.23).

Mazurek, M.J., Zielinski, W.J., 2004. "Individual legacy trees influence vertebrate wildlife diversity in commercial forests." *Forest Ecology and Management* 193, 321–334. https://doi.org/10.1016/j.foreco.2004.01.013

McAvoy, W.A., Harrison, J.W., 2012. "Plant community classification and the flora of Native American shell-middens on the Delmarva peninsula." *The Maryland Naturalist* 52, 1–34.

McBride, J.R., Jacobs, D.F., 1986. "Presettlement forest structure as a factor in urban forest development." *Urban Ecology* 9, 245–266.

McCarthy, H., 1993. "Managing oaks and the acorn crop," in: Blackburn, T.C., Anderson, M.K. (Eds.), *Before the Wilderness: Environmental Management by Native Californians*. Banning: Ballena Press, pp. 213–228.

McClaran, M.P., Bartolome, J.W., 1989. "Fire-related recruitment in stagnant *Quercus douglasii* populations." *Canadian Journal of Forest Research* 19, 580–585.

McCormick, A.C., Irmisch, S., Boeckler, G.A., Gershenzon, J., Köllner, T.G., Unsicker, S.B., 2019. "Herbivore-induced volatile emission from old-growth black poplar trees under field conditions." *Scientific Reports* 9, 7714. https://doi.org/10.1038/s41598-019-43931-y

McGregor, J.C., 1936. "The effects of a volcanic cinder fall on tree growth." *Tree Ring Bulletin* 3, 11–13.

Mensing, S., 2005. "The history of oak woodlands in California, Part I: The paleoecologic record." *The California Geographer* 45, 1–38.

Metz, M.R., Varner, J.M., Frangioso, K.M., Meentemeyer, R.K., Rizzo, D., 2012. "Collateral damage: Fire and *Phytophthora ramorum* interact to increase mortality in coast redwood." *Proceedings of the Sudden Oak Death Fifth Science Symposium USDA General Technical Report* PSW-GTR-243, 65–66.

Metz, M.R., Frangioso, K.M., Meentemeyer, R.K., Rizzo, D.M., 2011. "Interacting disturbances: wildfire severity affected by stage of forest disease invasion." *Ecological Applications* 21, 313–320. https://doi.org/10.1890/10-0419.1

Michel, P., Burritt, D.J., Lee, W.G., 2011. "Bryophytes display allelopathic interactions with tree species in native forest ecosystems." *Oikos* 120, 1272–1280.

Moritz, M.A., Odion, D.C., 2005. "Examining the strength and possible causes of the relationship between fire history and Sudden Oak Death." *Oecologia* 144, 106–114. https://doi.org/10.1007/s00442-005-0028-1

Muir, J., 1915. *Travels in Alaska*. New York: Houghton Mifflin Company.

Mulder, P.G., 2017. "Monitoring adult weevil populations in pecan and fruit trees in Oklahoma" [WWW Document]. Oklahoma State University. URL https://extension.okstate.edu/fact-sheets/monitoring-adult-weevil-populations-in-pecan-and-fruit-trees-in-oklahoma.html (accessed 5.10.23).

BIBLIOGRAPHY

Nelson, L.-A., Sanborn, P., Cade-Menun, B.J., Walker, I.J., Lian, O.B., 2021. "Pedological trends and implications for forest productivity in a Holocene soil chronosequence, Calvert Island, British Columbia, Canada." *Canadian Journal of Soil Science* 101, 654–672. https://doi.org/10.1139/cjss-2021-0033

Nelson, M.K., 2018. "Conclusion: Back in our tracks-embodying kinship as if the future mattered," in: Nelson, M.K., Schilling, D. (Eds.), *Traditional Ecological Knowledge: Learning from Indigenous Practices for Environmental Sustainability.* Cambridge: Cambridge University Press, pp. 250–266.

Nelson, M.K., Shilling, D. (Eds.), 2018. *Traditional Ecological Knowledge: Learning from Indigenous Practices for Environmental Sustainability.* Cambridge: Cambridge University Press.

Nelson, N., 1916. "Excavation of the Emeryville shellmound, 1906: Nels C. Nelson's final report." *Contributions of the University of California Archeological Research Facility* 54.

Noss, R., 1999. *The Redwood Forest: History, Ecology, and Conservation of the Coast Redwoods.* Washington DC: Island Press.

Odum, E.P., 1971. *Fundamentals of Ecology*, 3rd ed. Philadelphia: W.B. Saunders Co.

Olmsted, F.L., Vaughan, G.B., 1954. "A landscape of beauty and meaning," in: Drury, A. (Ed.), *Point Lobos Reserve: Interpretation of a Primitive Landscape.* State of California: Department of Natural Resources, pp. 19–36.

Oneal, C.B., Stuart, J.D., Steinberg, S.J., Fox, L., 2006. "Geographic analysis of natural fire rotation in the California redwood forest during the suppression era." *Fire Ecology* 2, 73–99. https://doi.org/10.4996/fireecology.0201073

Orús, M.I., Estévez, M.F., Vicente, C., 1981. "Manganese depletion in chloroplasts of *Quercus rotundifolia* during chemical simulation of lichen epiphytic states." *Physiologia Plantarum* 52, 263–266. https://doi.org/10.1111/j.1399-3054.1981.tb08503.x

Östlund, L., Zegers, G., Cáceres Murrie, B., Fernández, M., Carracedo-Recasens, R., Josefsson, T., Prieto, A., Roturier, S., 2020. "Culturally modified trees and forest structure at a Kawésqar ancient settlement at Río Batchelor, western Patagonia." *Human Ecology* 48, 585–597. https://doi.org/10.1007/s10745-020-00200-1

Ouimet, R., Duchesne, L., Moore, J.-D., 2017. "Response of northern hardwoods to experimental soil acidification and alkalinisation after 20 years." *Forest Ecology and Management* 400, 600–606. https://doi.org/10.1016/j.foreco.2017.06.051

Partlow, J., 2022. "California's giant sequoias are burning up. Will logging save them?" *The Washington Post* https://www.washingtonpost.com/climate-environment/2022/08/16/giant-sequoias-fire-mariposa-grove/.

Peattie, D.C., 1950. *A Natural History of Western Trees.* Lincoln: University of Nebraska Press.

Pisias, N.G., Mix, A.C., Heusser, L., 2001. "Millennial scale climate variability of the northeast Pacific Ocean and northwest North America based on radiolaria and pollen." *Quaternary Science Reviews* 20, 1561–1576. https://doi.org/10.1016/S0277-3791(01)00018-X

Pizňak, M., Bačkor, M., 2019. "Lichens affect boreal forest ecology and plant

BIBLIOGRAPHY

metabolism." *South African Journal of Botany* 124, 530–539. https://doi.org/10.1016/j.sajb.2019.06.025

Pizňak, M., Kolarčik, V., Goga, M., Bačkor, M., 2019. "Allelopathic effects of lichen metabolite usnic acid on growth and physiological responses of Norway spruce and Scots pine seedlings." *South African Journal of Botany* 124, 14–19. https://doi.org/10.1016/j.sajb.2019.04.011

Porada, P., Weber, B., Elbert, W., Pöschl, U., Kleidon, A., 2013. "Estimating global carbon uptake by lichens and bryophytes with a process-based model." *Biogeosciences* 10, 6989–7033. https://doi.org/10.5194/bg-10-6989-2013

Pyne, S., 2021. *The Pyrocene*. Oakland: University of California Press.

Radcliffe, D.C., Hix, D.M., Matthews, S.N., 2021. "Predisposing factors' effects on mortality of oak (*Quercus*) and hickory (*Carya*) species in mature forests undergoing mesophication in Appalachian Ohio." *Forest Ecosystems* 8, 1–14. https://doi.org/10.1186/s40663-021-00286-z

Reaves, J.L., Shaw, C.G., Mayfield, J.E., 1990. "The effects of *Trichoderma* spp. isolated from burned and non-burned forest soils on the growth and development of *Armillaria ostoyae* in culture." *Northwest Science* 64, 39–44.

Reid, C., Watmough, S.A., 2014. "Evaluating the effects of liming and wood-ash treatment on forest ecosystems through systematic meta-analysis." *Canadian Journal of Forest Research* 44, 867–885. https://doi.org/10.1139/cjfr-2013-0488

Rick, T., Ontiveros, M.Á.C., Jerardino, A., Mariotti, A., Méndez, C., Williams, A.N., 2020. "Human-environmental interactions in Mediterranean climate regions from the Pleistocene to the Anthropocene." *Anthropocene* 31, 100253. https://doi.org/10.1016/j.ancene.2020.100253

Rigg, G.B., 1918. "Growth of trees in *Sphagnum*." *Botanical Gazette* 65, 359–362.

Rogers, D.L., 2000. "Genotypic diversity and clone size in old-growth populations of coast redwood (*Sequoia sempervirens*)." *Canadian Journal of Botany* 78, 1408–1419. https://doi.org/10.1139/b00-114

Rozas, V., 2005. "Dendrochronology of pedunculate oak (*Quercus robur* L.) in an old-growth pollarded woodland in northern Spain: establishment patterns and the management history." *Annals of Forest Science* 62, 13–22. https://doi.org/10.1051/forest:2004091

Russo, M., 2014. "Ringed shell features of the southeastern United States," in: Roksandic, M., de Souza, S.M., Eggers, S., Burchell, M., Klokler, D. (Eds.), *The Cultural Dynamics of Shell-Matrix Sites*. Albuquerque: University of New Mexico Press, pp. 21–39.

Sarris, G., 2018. "The ancient ones," in: *The Once and Future Forest*. Berkeley: Heyday, pp. 101–127.

Sauer, C., 1952. *Agricultural Origins and Dispersals*. New York: American Geographical Society.

Saunders, R., 2014. "Shell rings of the lower Atlantic coast of the United States," in: Roksandic, M., Mendonca de Souza, S., Eggers, S., Burchell, M., Klokler, D. (Eds.), *The Cultural Dynamics of Shell-Matrix Sites*. Albuquerque: University of New Mexico Press, pp. 41–55.

BIBLIOGRAPHY

Schang, K., Cox, K., Trant, A.J., 2022. "Habitation sites influence tree community assemblages in the Great Bear rainforest, British Columbia, Canada." *Frontiers in Ecology and Evolution* 9, 791047. https://doi.org/10.3389/fevo.2021.791047

Schmidt, M.J., Goldberg, S.L., Heckenberger, M., Fausto, C., Franchetto, B., Watling, J., Lima, H., Moraes, B., Dorshow, W.B., Toney, J., Kuikuro, Yamalui, Waura, K., Kuikuro, H., Kuikuro, T.W., Kuikuro, T., Kuikuro, Yahila, Kuikuro, A., Teixeira, W., Rocha, B., Honorato, V., Tavares, H., Magalhães, M., Barbosa, C.A., Da Fonseca, J.A., Mendes, K., Alleoni, L.R.F., Cerri, C.E.P., Arroyo-Kalin, M., Neves, E., Perron, J.T., 2023. "Intentional creation of carbon-rich dark earth soils in the Amazon." *Science Advances* 9, eadh8499. https://doi.org/10.1126/sciadv.adh8499

Schriver, M., Sherriff, R.L., Varner, J.M., Quinn-Davidson, L., Valachovic, Y., 2018. "Age and stand structure of oak woodlands along a gradient of conifer encroachment in northwestern California." *Ecosphere* 9, e02446. https://doi.org/10.1002/ecs2.2446

Shao, S., Driscoll, C.T., Johnson, C.E., Fahey, T.J., Battles, J.J., Blum, J.D., 2016. "Long-term responses in soil solution and stream-water chemistry at Hubbard Brook after experimental addition of wollastonite." *Environmental Chemistry* 13, 528–540. https://doi.org/10.1071/EN15113

Shaul, D.L., 2019. *Esselen Studies: Language, Culture, and Prehistory*. Munich: LINCOM GmbH.

Shipek, F.C., 2014. "An example of intensive plant husbandry: The Kumeyaay of southern California," in: Harris, D.R., Hillman, G.C. (Eds.), *Foraging and Farming*. London: Routledge, pp. 159–170.

Shive, K.L., Wuenschel, A., Hardlund, L.J., Morris, S., Meyer, M.D., Hood, S.M., 2022. "Ancient trees and modern wildfires: Declining resilience to wildfire in the highly fire-adapted giant sequoia." *Forest Ecology and Management* 511, 120110. https://doi.org/10.1016/j.foreco.2022.120110

Sillett, S.C., Van Pelt, R., 2007. "Trunk reiteration promotes epiphytes and water storage in an old-growth redwood forest canopy." *Ecological Monographs* 77, 335–359. https://doi.org/10.1890/06-0994.1

Simard, S., 2021. *Finding the Mother Tree: Uncovering the Wisdom and Intelligence of the Forest*. New York: Vintage Books.

Singleton Hyde, H.M., Wallis, N.J., 2022. "Coastal subsistence and platform mound feasting on Florida's northern Gulf Coast." *The Journal of Island and Coastal Archaeology* 17, 375–401. https://doi.org/10.1080/15564894.2020.1783037

Smith, C.K., McGrath, D.A., 2011. "The alteration of soil chemistry through shell deposition on a Georgia (U.S.A.) barrier island." *Journal of Coastal Research* 27, 103–109. https://doi.org/10.2112/JCOASTRES-D-09-00086.1

Smith, D.J., 2010. "The history of temperate agroforestry." *Progressive Farming Trust Limited* 1–17.

Smith, H.I., 1903. "Part IV — Shell-heaps of the lower Fraser River, British Columbia." *Memoirs of the American Museum of Natural History* 4, 133–192.

BIBLIOGRAPHY

Smith, J., 1898. "The peach borer (*Sanninoidea exitiosa* Say.), experiments with hydraulic cement." *New Jersey Agricultural Experiment Stations Bulletin* 128, 1–28.

Smith, J.M.B., Klinger, L.F., 1985. "Aboveground-belowground phytomass ratios in Venezuelan paramo vegetation and their significance." *Arctic and Alpine Research* 17, 189–198. https://doi.org/10.2307/1550848

Soudzilovskaia, N.A., Graae, B.J., Douma, J.C., Grau, O., Milbau, A., Shevtsova, A., Wolters, L., Cornelissen, J.H.C., 2011. "How do bryophytes govern generative recruitment of vascular plants?" *New Phytologist* 190, 1019–1031. https://doi.org/10.1111/j.1469-8137.2011.03644.x

Stahle, D.W., Griffin, R.D., Meko, D.M., Therrell, M.D., Edmondson, J.R., Cleaveland, M.K., Stahle, L.N., Burnette, D.J., Abatzoglou, J.T., Redmond, K.T., 2013. "The ancient blue oak woodlands of California: Longevity and hydroclimatic history." *Earth Interactions* 17, 1–23.

Stebbins, G.L., Major, J., 1965. "Endemism and speciation in the California flora." *Ecological Monographs* 35, 1–35. https://doi.org/10.2307/1942216

Steel, Z.L., Goodwin, M.J., Meyer, M.D., Fricker, G.A., Zald, H.S.J., Hurteau, M.D., North, M.P., 2021. "Do forest fuel reduction treatments confer resistance to beetle infestation and drought mortality?" *Ecosphere* 12, e03344. https://doi.org/10.1002/ecs2.3344

Stephens, S.L., Kane, J.M., Stuart, J.D., 2018. "North coast bioregion" in: Van Wagtendonk, J.W., Sugihara, N.G., Stephens, Scott L., Thode, A.E., Shaffer, K.E., Fites-Kaufman, J.A. (Eds.), *Fire in California's Ecosystems*. Oakland: University of California Press, pp. 149–170.

Stephens, S.L., Fry, D.L., 2005. "Fire history in coast redwood stands in the northeastern Santa Cruz Mountains, California." *Fire Ecology* 1, 2-19. https://doi.org/10.4996/fireecology.0101002

Stephens, S.L., Martin, R.E., Clinton, N.E., 2007. "Prehistoric fire area and emissions from California's forests, woodlands, shrublands, and grasslands." *Forest Ecology and Management* 251, 205–216. https://doi.org/10.1016/j.foreco.2007.06.005

Stewart, O.C., 2002. *Forgotten Fires: Native Americans and the Transient Wilderness*. Norman: University of Oklahoma Press.

Stillitoe, P. (Ed.), 2017. *Indigenous Knowledge: Enhancing its Contribution to Natural Resources Management*. Boston: CABI.

Taliaferro, S., 2015. "Documentation and testing of nineteenth-century limewash recipes in the United States." (M.Sc. diss.), Columbia University, New York.

Thomas, S.C., Gale, N., 2015. "Biochar and forest restoration: a review and meta-analysis of tree growth responses." *New Forests* 46, 931–946. https://doi.org/10.1007/s11056-015-9491-7

Trant, A.J., Nijland, W., Hoffman, K.M., Mathews, D.L., McLaren, D., Nelson, T.A., Starzomski, B.M., 2016. "Intertidal resource use over millennia enhances forest productivity." *Nature Communications* 7, 12491. https://doi.org/10.1038/ncomms12491

Trowbridge, A.M., Stoy, P.C., 2013. "BVOC-mediated plant-herbivore interactions," in: Niinemets, Ü., Monson, R.K. (Eds.), *Biology, Controls and Models of Tree Volatile Organic Compound Emissions*. Dordrecht: Springer, pp. 21–46. https://doi.org/10.1007/978-94-007-6606-8_2

Tscharntke, T., Thiessen, S., Dolch, R., Boland, W., 2001. "Herbivory, induced resistance, and interplant signal transfer in *Alnus glutinosa*." *Biochemical Systematics and Ecology* 29, 1025–1047. https://doi.org/10.1016/S0305-1978(01)00048-5

Turner, N.J., Ari, Y., Berkes, F., Davidson-Hunt, I., Ertug, Z.F., Miller, A., 2009. "Cultural management of living trees: An international perspective." *Journal of Ethnobiology* 29, 237–270. https://doi.org/10.2993/0278-0771-29.2.237

Úbeda, X., Outeiro, L.R., 2009. "Physical and chemical effects of fire on soil," in: Cerdà, A., Robichaud, P.R. (Eds.), *Fire Effects on Soils and Restoration Strategies*. Boca Raton: CRC Press, pp. 121–148.

Ulery, A.L., Graham, R.C., Amrhein, C., 1993. "Wood-ash composition and soil pH following intense burning." *Soil Science* 156, 358–364.

Vadeboncoeur, M.A., 2010. "Meta-analysis of fertilization experiments indicates multiple limiting nutrients in northeastern deciduous forests." *Canadian Journal of Forest Research* 40, 1766–1780. https://doi.org/10.1139/X10-127

van Tooren, B.F., 1990. "Effects of a bryophyte layer on the emergence of seedlings of chalk grassland species." *Acta Oecologica* 11, 155-l63.

Vanderplank, S.E., Mata, S., Ezcurra, E., 2014. "Biodiversity and archeological conservation connected: Aragonite shell middens increase plant diversity." *BioScience* 64, 202–209. https://doi.org/10.1093/biosci/bit038

Veblen, T.T., Lorenz, D.C., 1991. *The Colorado Front Range: A Century of Ecological Change*. Salt Lake City: University of Utah Press.

Vierling, L.A., 1999. "Light heterogeneity and gas exchange dynamics above and within a monodominant Congolese rain forest canopy." (Ph.D. diss.), University of Colorado, Boulder.

Voelker, S.L., Merschel, A.G., Meinzer, F.C., Ulrich, D.E.M., Spies, T.A., Still, C.J., 2019. "Fire deficits have increased drought sensitivity in dry conifer forests: Fire frequency and tree-ring carbon isotope evidence from Central Oregon." *Global Change Biology* 25, 1247–1262. https://doi.org/10.1111/gcb.14543

Wardle, D.A., Walker, L.R., Bardgett, R.D., 2004. "Ecosystem properties and forest decline in contrasting long-term chronosequences." *Science* 305, 509–513. https://doi.org/10.1126/science.1098778

Weber, D., Seitfrit, D., 2023. "*Cytospora* canker of stone fruits in the home fruit planting." [WWW Document]. Penn State Extension. URL https://extension.psu.edu/cytospora-canker-of-stone-fruits-in-the-home-fruit-planting (accessed 5.10.23).

Whitehead, J., Wittemann, M., Cronberg, N., 2018. "Allelopathy in bryophytes — a review." *Lindbergia* 41, 1–7. https://doi.org/10.25227/linbg.01097

Wildcat, D.R., 2023. *On Indigenuity: Learning the Lessons of Mother Earth*. Wheat Ridge: Fulcrum Publishing.

Woglum, R.S., Lewis, H.C., 1940. "Whitewash to control potato leafhopper on

BIBLIOGRAPHY

citrus." *Journal of Economic Entomology* 33, 83–85. https://doi.org/10.1093/jee/33.1.83

Wohlleben, P., 2016. *The Hidden Life of Trees*. Vancouver: Greystone Books.

Xu, H., Cai, A., Wu, D., Liang, G., Xiao, J., Xu, M., Colinet, G., Zhang, W., 2021. "Effects of biochar application on crop productivity, soil carbon sequestration, and global warming potential controlled by biochar C:N ratio and soil pH: A global meta-analysis." *Soil and Tillage Research* 213, 105125. https://doi.org/10.1016/j.still.2021.105125

Young, H., 2021. "What can the marine shell remains tell us about diet, resource exploitation, function and ritual at the Cairns during the Iron Age?" (M.Sc. diss.), University of the Highlands and Islands, Inverness, Scotland.

Younging, G., 2018. *Elements of Indigenous Style: A Guide for Writing By and About Indigenous Peoples*. Canada: Brush Education.

Zeller, S.M., 1936. "Physical injuries to trees, with special reference to winter injury." *Oregon State Agricultural College Extension Service* 485, 1–8.

Zermeño, A., Gil, J.A., Hernández, A., Rodríguez, R., Ramírez, H., Benavides, A., Jasso, D., Munguia, J., Ibarra, L., 2010. "Effect of the entire whitewashing and TDZ application on budbreak, yield and quality of apple cv. Golden Delicious." *Bioagro* 22, 75–80.

ABOUT THE AUTHOR

Lee Klinger, Ph.D., is an Independent Scientist and Consultant in Big Sur, CA currently working with the Department of Natural Resources of the Esselen Tribe of Monterey County, and with the Mutsun Costanoan leaders at Indian Canyon Nation. Since 2005 he has served as the director of Sudden Oak Life, a movement aimed at applying fire mimicry practices to address the problems of forest decline and severe wildfires in California. He has more than forty years of experience in forestry, plant and soil ecology, atmospheric chemistry, earth system science, and nature photography, and has held scholarly appointments at the National Center for Atmospheric Research, the University of Colorado, the University of Oxford, the Chinese Academy of Sciences, and the Geological Society of London.

NOTES

PREFACE AND ACKNOWLEDGMENTS

1. Kroeber (1907).
2. Breschini and Haversat (2004), p. ii. Alternatively, Shaul (2019) believes that the Esselen name came from the Excelen, a local Esselen tribe of the upper Carmel Valley.
3. Kroeber (1925).
4. https://www.essellentribe.org
5. Breschini and Haversat (2004), pp. 62-66; Shaul (2019).
6. Https://www.esseltribe.org/history
7. Kimmerer (2014).
8. John Muir was a well-known naturalist of the late nineteenth and early twentieth centuries. His writings have inspired the preservationist environmental movement which defines "wilderness" as any place untouched by humans. He never once alluded to the ecological importance of the Native People in creating the wilderness. After reading his book *Travels in Alaska* (Muir 1915), it also bothered me that he frequently demeaned the Native People who were paddling his boat.
9. Lovelock (1995).
10. Klinger (1992); Klinger et al. (1996); Klinger and Erickson (1997); Klinger (2004).
11. Https://soars.ucar.edu/
12. Younging (2018).

PROLOGUE

1. Https://www.ksbw.com/article/cause-of-big-sur-pfeiffer-fire-released-by-national-forest-service/1055486.
2. For a detailed first-hand account of the Sobranes fire see T. Maehr (2024).
3. Https://suddenoaklifeorg.wordpress.com/2020/09/20/tom-little-bear-nason-esselen-elder-on-the-history-of-fire-management-in-big-sur/.

CHAPTER 1

1. Kimmerer (2003), p. vii.
2. Cajete (2006), p. 248.
3. Stillitoe (2017).
4. Long et al. (2020).

NOTES

5. Kimmerer (2013).
6. Martinez (2018), p. 140.
7. Nelson (2018), p. 253.
8. Anderson (2005).
9. Kimmerer (2014).
10. Kimmerer and Lake (2001).
11. Ellis et al. (2021).

CHAPTER 2

1. Heusser (1998); Pisias et al. (2001); Langenheim and Durham (1963).
2. Mensing (2005).
3. Vizcaino Expedition (1602-03) pp. 91-92; https://www.americanjourneys.org/aj-002/index.asp.
4. Vizcaino Expedition (1602-03) p. 91; https://www.americanjourneys.org/aj-002/index.asp.
5. Https://en.wikipedia.org/wiki/Vizca%C3%ADno-Serra_Oak.
6. The species taxonomy for common names used here is shown in the Appendix at the end of this book.
7. Most of the pollarded oaks I have seen are greater than ~36 inches diameter at breast height (dbh). In examining published studies on dbh-age relationships of oaks in California, findings indicate that for black oaks, Oregon white oaks, and blue oaks trees with a dbh greater than 36 inches mostly date from pre-settlement times (McClaran and Bartolome 1989; Garrison et al. 2002; Stahle et al. 2013; Schriver et al. 2018). In fact, Garrison et al. (2002) report that 17 percent of the black oak trees they dated from a central Sierra Nevada site were more than 200 years in age. Shriver et al. (2018) likewise report that between 5 and 10 percent of the sampled oaks in California's Northern Coast Ranges date from before 1850. While there are no studies reporting a dbh-age relationship for coast live oaks, there is evidence that many (~25 percent) of the larger coast live oaks in the San Francisco Bay area originated prior to the settlement era, approximately 1800 AD (McBride and Jacobs 1986). And, discussing blue oaks, Stahle et al. (2013) writes that "these stands also preserve an important fraction of trees that were recruited before the arrival of the first European settlers." The oldest living blue oak tree sampled during their study was at least 459 years old and many living trees were sampled in the 350 to 400-year age class. Several dead blue oaks also featured more than five hundred annual rings — with the oldest, located at Los Lobos, exhibiting 553 annual rings (indicating a lifespan from ca. AD 1333 to 1884). Based on these results as well as age data of young blue oaks (Koenig and Knops 2007), it is reasonable to conclude that some surviving blue oaks in California are more than six hundred years old, and that a very few select individuals might be as much as one thousand years old. Indeed, blue oaks are among the oldest angiosperms ever documented with exact tree-ring dating (Stahle et al. 2013), and it is reasonable to conclude that tens of thousands of

NOTES

ancient blue oaks and other species of oaks that were present before colonization still survive across the foothills of the Coast Ranges and Sierra Nevada.
8. Turner et al. (2009).
9. Fay (undated).
10. Fay (undated).
11. Mansion (undated).
12. Rozas (2005).
13. Asouti and Kabukcu (2014).
14. Smith (2010).
15. Greenlee and Langenheim (1990).
16. McCarthy (1993), Long et al. (2017).
17. Turner et al. (2009), p. 237.
18. Helen McCarthy quoted in Breschini and Haversat (2004), p 121.
19. All Esselen language translations are from Shaul (2019).
20. Long et al. (2017).
21. Bowcutt (2013).
22. Anderson (2005).
23. Lightfoot and Parrish (2009).
24. Marianchild (2014).
25. Hipp et al. (2018).
26. Shipek (2014).
27. Masters (2015).
28. Peattie (1950), p. 451.
29. Arneth et al. (2007).
30. Delmas et al. (1999).
31. Dr. Lee Vierling is currently an Associate Dean of Research and University Distinguished Professor, Department of Natural Resources and Society, University of Idaho, Moscow, ID.
32. Vierling (1999).
33. In most trees, light saturation, the point at which more sunlight no longer increases photosynthesis, occurs at about 10 percent of full sunlight for shade (understory) leaves and 50 to 70 percent for sun (upper canopy) leaves. This means that full sunlight is not necessary to achieve maximum rates of photosynthesis in a forest ecosystem.
34. Helmig et al. (1999a); Helmig et al. (1999b).
35. Tscharntke et al. (2001); Baldwin et al. (2006); Trowbridge and Stoy (2013); McCormick et al. (2019).
36. Koenig and Knops (2005).
37. Simard (2021).
38. Wohlleben (2016).
39. Karst et al. (2023).

NOTES

CHAPTER 3

1. Sillett and Van Pelt (2007); Lowman (2018).
2. Mazurek and Zielinski (2004).
3. Douhovnikoff et al. (2004).
4. Rogers (2000); Douhovnikoff et al. (2004).
5. The oldest redwood tree recorded in this region is nearly 1,300 years, as reported at: https://bigsurlandtrust.org/mitteldorf-preserve-carmel-valley/oldest-known-redwood/.
6. Stephens and Fry (2005).
7. Greenlee and Langenheim (1990).
8. Stephens et al. (2018).
9. Lorimer et al. (2009); Keeley (1982).
10. Oneal et al. (2006).
11. Brown and Baxter (2003).
12. Lutz et al., (2009).
13. Lightfoot and Parrrish (2009), pp. 198-199, p. 225; Sarris (2018), pp. 101-102.
14. Sarris (2018), p. 101.
15. Anderson (2005), p. 58.
16. Peattie (1950), p. 21.
17. Anderson (2005), p. 76.
18. Noss (1999), quoted from https://en.wikipedia.org/wiki/Sequoia_sempervirens.
19. Stephens et al. (2018), p. 158.
20. Henrikson (2017).

CHAPTER 4

1. Jepson (1954a).
2. Stebbins and Major (1965).
3. Johnson (1977); Axelrod (1982).
4. While the focus here is on the origin and distribution of Monterey cypresses, many of these points are equally applicable to the Monterey pines.
5. Peattie (1950), p. 241.
6. The earliest record of local propagation of Monterey cypresses was in 1876 at a nursery near Monterey, as cited by Gordon (1974).
7. Drury and Neasham (1954).
8. Olmsted and Vaughan (1954); Drury (1954).
9. Late in developing or opening.
10. Peattie (1950), p. 244. It also seems relevant that the Monterey pine was not described until 1836. I'm puzzled by the fact that the early botanists missed categorizing this prominent tree species as well!
11. The Gymnosperm Database: https://www.conifers.org/cu/Cupressus_macrocarpa.php.
12. Hartweg (1847); http://www.cupressus.net/CUmacrocarpaHartweg.html.

NOTES

13. Hartweg (1847) stated in the description of the species that, due to hostilities brought on by the Mexican-American War, it was not safe for him to travel more than a few miles from Monterey in July of 1846, meaning that he visited only one (Del Monte forest) relic cypress population.
14. Jepson (1923) states that the occurrence of Monterey cypress was first reported by the La Perouse Expedition of 1786. Indeed, La Perouse (1807) mentions cypress twice, first in reference to a location "Point Cypress" (presumably near Pebble Beach), and second in reference to "the cypress" as being one of the nearby trees. Although La Perouse claims his botanists "did not lose a moment in adding to their collections," they somehow missed describing the Monterey cypress. While it is likely that previous explorers, including Vizcaino in 1602, saw and visited the cypresses at Point Lobos, as far as is known they are not mentioned explicitly in any expedition prior to La Perouse.
15. Griffin and Critchfield (1972); MacBride (1913).
16. Shaul (2019).
17. The latest (2023) draft of the *Big Sur Coast Land Use Plan* discourages the planting of and *encourages the removal of* the "non-native" Monterey cypress trees within the Big Sur Coastal Planning Area.
18. Jepson (1923), p. 75. With all due respect to W. Jepson, his statement that "the age of mature Monterey cypress is about 50 to 300 years" is mere hand-waving without actual age data. Peattie (1950) echoes Jepson's claims stating that that the largest cypresses in these groves are on average two centuries old, three at most, while adding that "little indeed is known about their ages."
19. Jepson (1954b).
20. MacBride and Froehlich (1984).

CHAPTER 5

1. Klinger (1988).
2. Jerardino (2016); Rick et al. (2020).
3. Https://en.wikipedia.org/wiki/Turtle_Mound.
4. Saunders (2014).
5. Klinger (2006).
6. Nelson (1916).
7. Https://en.wikipedia.org/wiki/Emeryville_Shellmound.
8. Nelson (1916), p. 10.
9. Klinger (2006), pp. 161-162.
10. Russo (2014).
11. Young (2021).
12. Singleton Hyde and Wallis (2022).
13. Smith (1903), p. 133.
14. Östlund et al. (2020).
15. Trant et al. (2016), p. 1.
16. Anderson (1952).
17. Doolittle (2000), p. 82.

NOTES

18. Doolittle (2000), p. 440.
19. Schmidt et al. (2023).
20. McAvoy and Harrison (2012).
21. Cook-Patton et al. (2014).
22. Vanderplank et al. (2014).
23. Fischer et al. (2019).
24. Schang et al. (2022).
25. Smith and McGrath (2011), p. 106.
26. Trant et al. (2016), p. 6. See also Hoffman et al. (2016) and Hoffman et al. (2017).
27. Armstrong et al. (2022); Fischer et al. (2019).
28. Harraz et al. (2020).
29. Before present (BP) in Western literature usually means "before 1950."
30. Jones (2003).

CHAPTER 6

1. Some bryophytes do have vascular tissue, but it is nonlignified and distinct from the vascular tissue of other plants.
2. Glime (2024).
3. Barrett and Arno (1999), p. 52.
4. Peattie (1950), p. 242.
5. Peattie (1950), p. 118.
6. Evernic acid ($C_{20}H_{18}O_8$) is a depside characterized by two phenolic hydroxyl groups linked by an ester bond, which exhibits antimicrobial properties, inhibiting the growth of bacteria and fungi, thus contributing to the lichen's defense against microbial pathogens. Usnic acid ($C_{18}H_{16}O_7$) is also a depside characterized by two phenolic hydroxyl groups linked by an ester bond with similar properties to evernic acid.
7. Legaz et al. (1988).
8. Orús et al. (1981); Ascaso and Rapsch (1986).
9. Legaz et al. (2004).
10. Pizňak and Bačkor (2019); Pizňak et al. (2019).
11. Glime (2024).
12. Glime (2024), p. 3.
13. Glime (2024), p. 43.
14. Harmon and Franklin (1989).
15. Glime (2024), p. 44.
16. Soudzilovskaia et al. (2011); see also van Tooren (1990).
17. Klinger (2005); Klinger (1990); Cornish (1999); Smith and Klinger (1985).
18. Klinger (1988).
19. Klinger (1996).
20. Rigg (1918).
21. Chiapusio et al. (2013); Michel et al. (2011); Whitehead et al. (2018).

NOTES

22. Klinger (1990); see also my preliminary nutrient analyses of moss vs. non-moss-covered soils around English oak trees affected by Acute Oak Decline showing lower soil fertility under mossy areas: https://suddenoaklifeorg.wordpress.com/2010/04/30/acute-oak-decline-in-the-uk-part-2/.
23. Glime (2024), p. 8.
24. Glime (2024).
25. Fritz and Brunet (2010); Glime (2024).
26. Glime (2024).
27. Klinger (2009); Klinger (1996); Klinger and Short (1996).
28. Klinger et al. (1983).
29. Glime (2024).
30. Clements (1916).
31. Odum (1971); Wardle (2004).
32. Klinger (1992); Klinger et al. (1996); Helbig et al. (2020); Porada et al. (2013).
33. Nelson et al. (2021).

CHAPTER 7

1. Voelker et al. (2018).
2. Wardle et al. (2004); Fenn et al. (2006); Radcliffe et al (2021).
3. Hauck (2003).
4. Huggett (1998).
5. Huntington et al. (2000); Hamburg et al. (2003). See also a summary of my rain pH measurements in Big Sur (2007-2012) revealing somewhat acidic values (pH 4.6-4.9), https://suddenoaklifeorg.wordpress.com/2013/02/03/acid-rain-in-big-sur-2011-2012-season-summary/.
6. Voelker et al. (2018).
7. Knight et al. (2022). Regarding factors driving the recent severe wildfires, I agree with Scott Stephens that "climate change is no more than 25% of the problem, [rather] it's 75% forest structure." https://www.ppic.org/blog/how-active-steward ship-could-protect-californias-forests-from-extreme-wildfire/.
8. Stephens et al. (2007).
9. Https://en.wikipedia.org/wiki/List_of_California_wildfires.
10. Shive et al. (2022).
11. Partlow (2022).
12. Folks, this is not normal! And please don't write it all off to climate change, then buy a Tesla. While climatology is certainly a factor here, so is ecology, and I believe that we can eventually solve this problem, at least regionally, through better forest stewardship involving lots of focused, physical work long before we can change the climate.
13. Anderson (2005), p. 145.
14. Close et al. (2009).
15. Jurskis (2009).
16. Greenler et al. (2024); see also Association for Fire Ecology, https://fireecology.org/.

NOTES

17. Leopold (1972).
18. Sauer (1952).
19. Stewart (2002).
20. Anderson (2002), p. 41.
21. Davis et al. (2024).
22. Hanson (2021), p. 25.
23. Crawford et al. (2015); Hoffman et al. (2016); Hoffman et al. (2017).
24. Pyne (2021).
25. Pyne (2021), p. 6.
26. Li et al. (2019).
27. Reid and Watmough (2014).
28. Vadeboncoeur (2010).
29. Li et al. (2014).
30. Https://hubbardbrook.org/wp-content/uploads/Watershed_1_CalciumAddition.pdf.
31. Shao et al. (2016); Battles et al. (2014); Johnson et al. (2014).
32. Hawley et al. (2006).
33. Juice et al. (2006).
34. Huggett et al. (2007).
35. Ouimet et al. (2017).
36. Bakker and Nys (1999).
37. Bakker et al. (2000).
38. Klinger (2005).
39. Bosman et al. (2001).
40. Úbeda and Outeiro (2009); Agbeshie et al. (2022); Ulery et al. (1993).
41. Ulery et al. (1993).
42. Eicher and Rounsefell (1957); McGregor (1936).
43. Fairhead et al. (2017).
44. Glaser and Birk (2012).
45. Downie et al. (2011); Fairhead et al. (2017).
46. Biederman and Harpole (2013); Xu et al. (2021); Thomas and Gale (2015).
47. Haider et al. (2022).
48. Moritz and Odion (2005).
49. Metz et al. (2011). The Composite Burn Index (CBI) is based on a ranking of burn effects on five different strata in the forest: substrate, herb, shrub, intermediate trees, and canopy trees. Here, however, the authors combine these CBI rankings with their own measurements of ash depth and height of canopy scorching at eight locations to arrive at a unique burn severity index.
50. Metz et al. (2011).
51. Beh et al. (2012).
52. Metz et al. (2012).
53. Reaves et al. (1990); Froelich et al. (1978).
54. Anderson et al. (1987).
55. Hood et al. (2016); Steel et al. (2021).
56. Anderson (2005).
57. Anderson (2006).

NOTES

58. McCarthy (1993), p. 221.
59. Halpern et al. (2022).
60. Aldern and Goode (2014).
61. Gallagher et al. (2022); Brennan et al. (2023).

CHAPTER 8

1. Video explaining fire mimicry: https://suddenoaklifeorg.wordpress.com/2023/07/13/video-what-is-fire-mimicry/.
2. Arno and Fiedler (2005).
3. Long (2009).
4. Let me say that I am very supportive of the shift from gas to electric chain saws and have just purchased my first battery-powered chain saw. This is a cleaner and more efficient technological advancement, allowing for quieter saw operation without the need to haul gas containers in the field. I like that I can simply throw a couple of batteries and some bar oil in my backpack and I'm set for the day.
5. Https://www.youtube.com/@leeklinger1088/videos.
6. Https://azomite.com/.
7. Https://azomite.com/about-azomite-mineral-products/history/.
8. Eudoxie and Martin (2019).
9. Taliaferro (2015).
10. Personally observed in an exhibit featuring historical photos of China at the Asian Art Museum of San Francisco in 2005.
11. Frantom (undated).
12. Research on limewash is quite lacking in the science of tree care. For my part, I have submitted *three* proposals to the U.S. Forest Service to study the effects of limewash on sick oaks, but to no avail.
13. Smith (1898). Smith's findings involve his use of "cement" as an ingredient of the trunk treatments. Cement is comprised mainly of hydrated lime.
14. Gusella et al. (2021).
15. Zermeño et al. (2010).
16. Gossard (1913).
17. Mulder (2017).
18. Zeller (1936).
19. Woglum and Lewis (1940).
20. Https://micronmetals.com/product/iron-oxide-powder-black-magnetite/.
21. This surgical approach was inspired, in part, by my dentist Dr. Roy Thomas who, while fixing a cavity in my tooth, told me how he removed bark beetles from his Monterey pine tree using a similar drilling procedure with power tools.
22. Weber and Seitfrit (2023); John Valenzuela, personal communication, 2022.
23. Axes should be of the type designed for cutting (narrow tapered head), not splitting (wide tapered head).
24. Https://www.youtube.com/@leeklinger1088/videos.

NOTES

CHAPTER 9

1. Veblen and Lorenz (1991).
2. Klinger (2008).
3. Https://en.wikipedia.org/wiki/Dolan_Fire.
4. Detailed case study results at www.suddenoaklife.org.

EPILOGUE

1. Martinez (2018), p. 154.
2. Https://suddenoaklifeorg.wordpress.com/2022/11/14/thoughts-and-images-from-our-restoring-fire-safe-communities-fire-mimicry-and-tek-workshop-at-indian-canyon-nov-11-13-2022/.
3. Https://ecocampcoyote.org/; https://santacruzpermaculture.com/; https://calpba.org/centralcoastpba/.
4. Wildcat (2023).
5. Levy (2022).
6. Lake and Christianson (2019); Lake et al. (2017).
7. Https://www.culturalfire.org/.
8. Martinez et al. (2023); see also https://www.culturalfire.org/projects.
9. Ellis et al. (2021); Mann (2011).

Made in the USA
Middletown, DE
21 May 2024

54621169R00115